Simon Hainz

Adaptive Predictive Signal Processing for Integrated Sensors

Simon Hainz

Adaptive Predictive Signal Processing for Integrated Sensors

Compensation of the Systematic Angle Error Caused by Air Gap Variations Using Decision Feedback Equalizer Approach

Südwestdeutscher Verlag für Hochschulschriften

Impressum/Imprint (nur für Deutschland/ only for Germany)
Bibliografische Information der Deutschen Nationalbibliothek: Die Deutsche Nationalbibliothek verzeichnet diese Publikation in der Deutschen Nationalbibliografie; detaillierte bibliografische Daten sind im Internet über http://dnb.d-nb.de abrufbar.
Alle in diesem Buch genannten Marken und Produktnamen unterliegen warenzeichen-, marken- oder patentrechtlichem Schutz bzw. sind Warenzeichen oder eingetragene Warenzeichen der jeweiligen Inhaber. Die Wiedergabe von Marken, Produktnamen, Gebrauchsnamen, Handelsnamen, Warenbezeichnungen u.s.w. in diesem Werk berechtigt auch ohne besondere Kennzeichnung nicht zu der Annahme, dass solche Namen im Sinne der Warenzeichen- und Markenschutzgesetzgebung als frei zu betrachten wären und daher von jedermann benutzt werden dürften.

Verlag: Südwestdeutscher Verlag für Hochschulschriften Aktiengesellschaft & Co. KG
Dudweiler Landstr. 99, 66123 Saarbrücken, Deutschland
Telefon +49 681 37 20 271-1, Telefax +49 681 37 20 271-0, Email: info@svh-verlag.de
Zugl.: Wien, Technische Universität, Dissertation, 2008

Herstellung in Deutschland:
Schaltungsdienst Lange o.H.G., Berlin
Books on Demand GmbH, Norderstedt
Reha GmbH, Saarbrücken
Amazon Distribution GmbH, Leipzig
ISBN: 978-3-8381-0827-8

Imprint (only for USA, GB)
Bibliographic information published by the Deutsche Nationalbibliothek: The Deutsche Nationalbibliothek lists this publication in the Deutsche Nationalbibliografie; detailed bibliographic data are available in the Internet at http://dnb.d-nb.de.
Any brand names and product names mentioned in this book are subject to trademark, brand or patent protection and are trademarks or registered trademarks of their respective holders. The use of brand names, product names, common names, trade names, product descriptions etc. even without a particular marking in this works is in no way to be construed to mean that such names may be regarded as unrestricted in respect of trademark and brand protection legislation and could thus be used by anyone.

Publisher:
Südwestdeutscher Verlag für Hochschulschriften Aktiengesellschaft & Co. KG
Dudweiler Landstr. 99, 66123 Saarbrücken, Germany
Phone +49 681 37 20 271-1, Fax +49 681 37 20 271-0, Email: info@svh-verlag.de

Copyright © 2009 by the author and Südwestdeutscher Verlag für Hochschulschriften Aktiengesellschaft & Co. KG and licensors
All rights reserved. Saarbrücken 2009

Printed in the U.S.A.
Printed in the U.K. by (see last page)
ISBN: 978-3-8381-0827-8

Kurzfassung

Das Messen des Winkels und der Geschwindigkeit von rotierenden Wellen ist eine der wichtigsten Aufgaben im Automobilbereich. Für derartige Messungen können verschiedene Messsysteme verwendet werden, die auf unterschiedlichen Verfahren (optisch, kapazitiv, induktiv, magnetisch, usw.) beruhen. Magnetische Sensoren sind sehr robust gegen äußere Einflüsse wie z.B. Verschmutzung durch Öl und Staub oder Temperaturvariationen. Wegen ihrer hohen Robustheit werden magnetische Sensoren vor allem in Motorapplikationen sehr häufig verwendet. Ein Nachteil magnetischer Winkelmesssysteme ist hingegen deren geringe Genauigkeit von $1°$ bis $2°$.

Ein in Motorapplikationen häufig verwendetes Winkelmesssystem besteht aus einer magnetisch kodierten Scheibe und einem Magnetfeldsensor. Die Scheibe wird an der rotierenden Welle montiert und der Sensor befindet sich an einer fixen Winkelposition. Dreht sich die Welle, so bewegt sich die Kodierung der Scheibe am Sensor vorbei. Das am Sensor auftretende Magnetfeld wird durch physikalische Effekte erfasst und in ein elektrisches Signal umgewandelt. Dieses elektrische Sensorsignal ist eine elektrische Abbildung der Kodierung der Scheibe. Ist dem Motorsteuergerät die Kodierung der Scheibe bekannt, so kann es durch das Auswerten des elektrischen Sensorsignals den Winkel und die Geschwindigkeit der Welle bestimmen.

Heutige Sensoren verwenden meist einen Nulldurchgangsdetektor, um die analoge Spannung in ein binäres Signal umzuwandeln. Der Nulldurchgang des analogen Signals kann sehr genau bestimmt werden und Winkelmessungen mit einer Genauigkeit von $0{,}06°$ sind möglich. Verändert sich jedoch der Luftspalt zwischen der kodierten Scheibe und dem Magnetfeldsensor, so kommt es zu einer Verschiebung der Nulldurchgänge und es entsteht dadurch ein systematischer Messfehler von bis zu $2°$.

In Motorapplikationen ändert sich der Luftspalt unter anderem durch mechanische Fertigungstoleranzen und Vibrationen. Fertigungstoleranzen können nicht verhindert werden. Vibrationen werden in Zukunft sogar zunehmen, da das Gewicht von Motoren weiter reduziert werden wird. Da Änderungen des Luftspalts auftreten, kann der systematische Winkelfehler nicht verhindert werden. Die Genauigkeit des Messsystems kann nur durch die Kompensation des systematischen Fehlers erhöht werden.

Durch neue Technologien von mikroelektronischen Schaltungen ist es heute möglich, auf relativ geringer Chipfläche bereits komplexe digitale Signalverarbeitung durchzuführen. Im Rahmen dieser Arbeit wurde untersucht, inwiefern digitale Signalverarbeitung eingesetzt werden kann, um den systematischen Winkelfehler zu kompensieren.

Die vorgestellte digitale Filterstruktur basiert auf dem so genannten »Decision Feedback Equalizer« (DFE), der vor allem in der digitalen Nachrichtentechnik verwendet wird. Der DFE wurde modifiziert und für die Anwendung in Messsystemen optimiert. Der modifizierte DFE kompensiert die Verschiebung der Nulldurchgänge und erhöht damit die Winkelgenauigkeit. Durch die Modifikation konnte die zur Realisierung benötigte Hardware, im Vergleich zum herkömmlichen DFE, stark reduziert werden.

Die Filterkoeffizienten des modifizierten DFEs werden durch einen adaptiven Algorithmus bestimmt. Die Filterkoeffizienten werden dabei nicht direkt, sondern über ein physikalisches Modell der Messanordnung adaptiert. Adaptiert man die physikalischen Parameterwerte anstelle der Filterkoeffizienten, so wird die Robustheit und Geschwindigkeit der Adaption erhöht. Zudem sind die physikalischen Parameterwerte (z.B. der Luftspalt) dem Algorithmus nach abgeschlossener Adaption bekannt und können für andere Zwecke verwendet werden.

Die Komplexität des modifizierten DFEs und adaptiven Algorithmus kann reduziert werden, indem man einen Systemtakt verwendet, der synchron zur Geschwindigkeit der rotierenden Welle ist. Die Nockenwelle wird durch die unterschiedlichen Phasen des Verbrennungsmotors (z.B. Kompression, Explosion) kurzzeitig beschleunigt oder abgebremst. Eine Phasenregelschleife (»phase locked loop«) kann diesen Geschwindigkeitsfluktuationen nicht folgen. Deshalb wird zur Erzeugung des Systemtakts eine Kombination aus einer Phasenregelschleife und einer variablen Verzögerungskette (»Delay Line«) eingesetzt. Die Phasenregelschleife erzeugt ein Signal mit konstanter Frequenz, welches durch die variable Verzögerungskette phasenmoduliert wird.

Durch die in dieser Arbeit vorgestellte Filterstruktur kann der systematische Winkelfehler des magnetischen Messsystems kompensiert werden. In den präsentierten Simulationsergebnissen wird der systematische Fehler von 1,8° auf 0,45° reduziert. Verwendet man einen DFE mit mehr Koeffizienten und höherer Taktfrequenz, so kann der Winkelfehler weiter reduziert werden. Die Genauigkeit des Messsystems wird durch das Filterdesign bestimmt.

Abstract

Precise speed and angle measurements of rotating shafts are very important in mechanical engineering applications, particularly in automotive engineering. Several different sensing techniques including optical, capacitive, inductive, and magnetic technique can be used to measure the angle and speed of a rotating shaft. Because of the harsh environment within the engine compartment, a very robust sensing arrangement is required for measurements. Due to their high degree of robustness and low production costs, magnetic field sensors have been the preferred types of sensors for engine applications in recent decades. The main disadvantage of using the magnetic technique for angle measurement is its limited angle accuracy, which ranges from 1° to 2°.

The typical sensing arrangement in engine applications consists of a magnetically patterned wheel that is mounted on a rotating shaft, and a magnetic field sensor element mounted at a fixed angular position. The sensor element converts the applied field into an electrical signal, which upon first consideration is linear in proportion to the applied magnetic field. The output signal of the sensor element is an electric image of the pattern on the wheel. If the engine control unit knows the magnetic pattern on the wheel, it can determine the speed and angle of the rotating shaft by evaluating the electric image.

Sensor concepts employed today use zero crossing detection to transform the analog electrical signal into a binary sequence. The accuracy of zero crossing detection is high and allows angle measurements with an accuracy of 0.06°. However, variations of the air gap between the sensor element and pole wheel cause displacements of the zero crossing positions and can result in a systematic angle error of up to 2°.

The systematic angle error can be avoided by keeping the air gap constant. However, mounting and packaging tolerances, as well as mechanical vibrations, cause air gap variations. Tolerances cannot be avoided, and vibrations are expected to increase in the future due to the reduction of engine weight. Therefore, in this application, the systematic angle error cannot be avoided and must be compensated.

High integration densities of new semiconductor technologies allow the integration of complex digital signal processing onto a small silicon area. Due to new technologies, high processing power is also available on the small chip area of integrated sensors.

The aim of this thesis is to find a solution for increasing the angle accuracy by compensating the gap-dependent displacement, using the processing power available on today's integrated magnetic field sensors.

This thesis presents one possible solution for overcoming the problem. The decision feedback equalizer (DFE), commonly used in digital communication, was modified for use in the sensing arrangement. The modified DFE is used to compensate the displacement, which improves the angle accuracy. The required hardware for the modified DFE is small compared to the conventional DFE.

An adaptive algorithm estimates the filter coefficients of the DFE. Rather than estimating the filter coefficients directly, a simple physical model of the sensing arrangement is used. The estimation of the physical parameter values allows an increase in the adaption speed and a reduction in the complexity of the adaptive algorithm. As a further benefit, the physical parameter values (e.g., the air gap) are estimated and become known after adaption is completed.

The complexity of the DFE and adaptive algorithm can be reduced if the clock signal is synchronous to the rotational speed of the shaft. Therefore, a modified phase locked loop (PLL) is used to generate a clock signal with variable frequency.

Simulations show that the displacement of the zero crossings can be compensated, and the angle accuracy can be reduced from 1.8° to 0.45°. Further improvements are possible by increasing the filter length and clock frequency. The accuracy of the sensing arrangement can now be defined by design.

Acknowledgement

First of all, I want to thank my first supervisor Prof. Herbert Grünbacher for his great support over the past few years. He encouraged me to undertake my doctoral studies at the Vienna University of Technology, gave me important feedback on technical problems, and often asked the right questions. I also want to thank him for his time and energy spent reading and correcting my thesis. With his help, I greatly improved my English and writing skills. Special appreciation also goes to Dirk Hammerschmidt, from Infineon Technologies Austria AG, for his fruitful ideas and helpful suggestions during discussions. He is a great technical advisor and I learned a lot from him. I would also like to thank Prof. Christoph Grimm, my second supervisor from the Vienna University of Technology, for his time and effort in evaluating and reviewing my thesis. Last but not least, I am grateful to Erwin Ofner, from Carinthia University of Applied Sciences, for his support during the first years of this work.

I want to acknowledge not only my supervisors, but also my colleagues who contributed to this work. Mario Jungwirth, from the Upper Austria University of Applied Sciences, supported me with FEM simulations. Tobias Werth, Mario Motz and David Tatschl, from Infineon Technologies Austria AG, discussed with me about results and possible improvements. Special thanks go also to Vladimír Košel, from KAI Kompetenzzentrum Automobil- und Industrie-Elektronik GmbH, for helping me to improve my writing skills.

My gratitude extends to all other colleagues from Infineon Technologies Austria AG and KAI Kompetenzzentrum Automobil- und Industrie-Elektronik. I thank them all for their helpful hints, and for the great working atmosphere, as well!

Many thanks also go to my parents, brothers and sister, my friends, and especially my girlfriend, Karoline, for their understanding that I was always pinched for time.

Finally, I want to acknowledge the financial support from »KAI Kompetenzzentrum Automobil- und Industrie-Elektronik GmbH,« »Infineon Technologies Austria AG,« »Carinthia University of Applied Sciences,« and the funding of the »Federal Ministry of Economics and Labour of the Republic of Austria« and the »Carinthian Economic Promotion Fund« (KWF) (contract 98.362/0112-C1/10/2005).

Table of Contents

CHAPTER 1 .. INTRODUCTION 1

CHAPTER 2 .. A BRIEF REVIEW OF FUNDAMENTALS 3
2.1 Decision Feedback Equalizer Fundamentals 3
    2.1.1 Linear Equalizer 3
    2.1.2 Decision Feedback Equalizer 4
    2.1.3 Partial Response Technique and Decision Feedback Equalizer ...6
    2.1.4 Adaptive Decision Feedback Equalizer.......................................7
2.2 Adaptive Algorithm Fundamentals 8
    2.2.1 Performance Function 9
    2.2.2 Adaptive Methods..10
    2.2.3 Learning Function 13
    2.2.4 Adapting Multiple Parameters 14
2.3 Phase Locked Loop Fundamentals 15
    2.3.1 Digital Phase Locked Loop 15
    2.3.2 All-Digital Phase Locked Loop...16

CHAPTER 3 .. SPEED AND ANGLE MEASUREMENTS IN AUTOMOTIVE APPLIC....19
3.1 Measurement Techniques 19
    3.1.1 Potentiometers 20
    3.1.2 Rotating Permanent Magnet 20
    3.1.3 Resolver...20
    3.1.4 Rotary Encoder..21
3.2 Engine Application 23
    3.2.1 Misfire Detection...24
    3.2.2 Camless Engine 24
3.3 Sensing Arrangements for Angle Measurement in Engine Application ...25
    3.3.1 Pole Wheel...25
    3.3.2 Toothed Wheel...27
    3.3.3 Magnetic Field Sensor Element...29

CHAPTER 4.. UNTREATED PROBLEM 31
4.1 Sensor Concepts in Use Today 31
    4.1.1 Estimation of the DC Component 32
    4.1.2 Conversion of the Sinusoidal-Like Waveform in a Binary Sequence 33
4.2 Untreated Problem of Sensor Concepts in Use Today 34
4.3 Reason for Research 35

CHAPTER 5.. SOLUTION OF THE PROBLEM 37
5.1 Introduction 37
    5.1.1 Reason for the Displacement of the Zero Crossing Positions 37
    5.1.2 Intersymbol Interference in Other Applications 41
    5.1.3 Known Techniques to Deal with ISI 42
    5.1.4 Aim of the Thesis 44
5.2 Displacement Compensation 45
    5.2.1 Increasing Angle Accuracy 45
    5.2.2 Predictive Feedback Filter 46
    5.2.3 Start-Up Phase 48
    5.2.4 Implementation Hints 49
    5.2.5 Complexity Estimation 51
5.3 Adaptive Algorithm using Physical Model of Sensing Arrangement 51
    5.3.1 Estimation of the Physical Parameter Values 53
    5.3.2 Simple Physical Model of Sensing Arrangement using Pole Wheel 54
    5.3.3 Error Signal 57
    5.3.4 Performance Function and Learning Function 59
    5.3.5 Complexity Estimation 61
5.4 Clock Signal Generation 61
    5.4.1 Constant Number of Samples per Symbol 61
    5.4.2 Crankshaft Speed Variations 65
    5.4.3 Crankshaft Speed Fluctuations 68
    5.4.4 Implementation Hints 74
    5.4.5 Complexity Estimation 77
5.5 pDFE with Adaptive Algorithm and PLL 77
    5.5.1 Simulation Results in the Time Domain 77
    5.5.2 Performance Limits 81

CHAPTER 6.. CONCLUSION 86
6.1 Summary 86
6.2 Outlook 88

List of Figures

Figure 2.1: Linear equalizer $g_r[c]$ 4
Figure 2.2: Linear equalizer realized as a filter in the feedback path.................................... 5
Figure 2.3: Intersymbol interference (pre-ISI, Symbol and post-ISI)..................................... 5
Figure 2.4: Decision feedback equalizer 6
Figure 2.5: Partial response equalizer.. 7
Figure 2.6: Partial response decision feedback equalizer 7
Figure 2.7: Adaptive decision feedback equalizer 8
Figure 2.8: Performance function of the number guessing problem................................. 10
Figure 2.9: Illustration of adaptive methods 12
Figure 2.10: Illustration of the simplex method.. 13
Figure 2.11: Learning function using different adaptive methods 14
Figure 2.12: Digital charge pump phase locked loop 15
Figure 2.13: Phase and frequency detector and current pump 16
Figure 2.14: All-digital phase locked loop... 16
Figure 2.15: Transfer characteristic of the PFD followed by the TDC............................... 17
Figure 2.16: Second order biquad IIR filter... 17
Figure 3.1: Angular speed of engine crankshaft at... 23
Figure 3.2: Sensing arrangement for angle measurement using pole wheel 26
Figure 3.3: Magnetic field of a pole wheel (FEM simulations) 27
Figure 3.4: Sensing arrangement for angle measurement using toothed wheel 28
Figure 3.5: Magnetic field of a back-bias magnet, influenced by a toothed wheel (FEM).......... 29
Figure 4.1: Commonly used sensor concepts 32
Figure 4.2: Magnetic field of a pole wheel (FEM simulations) 34
Figure 4.3: Magnetic field of a back-bias magnet, influenced by a toothed wheel (FEM).......... 35
Figure 5.1: Channel of the differential sensing arrangement with 38
Figure 5.2: Pulse shaped magnetic field generated by a single magnetic region 38
Figure 5.3: Magnetic field distribution generated by three magnetic regions 39
Figure 5.4: Magnetic field of a pole wheel (linear superposition)...................................... 40
Figure 5.5: General baseband transmission channel... 42
Figure 5.6: Transmission channel of the sensing arrangement in the time domain 43
Figure 5.7: Block diagram of filter structure... 45
Figure 5.8: Magnetic pattern along the circumference of the pole wheel 46
Figure 5.9: Predictive decision feedback equalizer... 47
Figure 5.10: Simulation result of the pDFE for a GMR sensor element 48
Figure 5.11: pDFE with start-up circuit 49
Figure 5.12: Replacement of the post-FBF by a counter and LUT with $\psi = 8$ 50

Figure 5.13: Replacement of the post-FBF by multiple counters and LUTs with $\psi = 8$ 50
Figure 5.14: pDFE with adaptive algorithm 52
Figure 5.15: Adaptive algorithm of the DFE and pDFE 53
Figure 5.16: Cross section of a permanent magnet 54
Figure 5.17: Simple model of a permanent magnet 55
Figure 5.18: Wire model of a magnetic strip 56
Figure 5.19: Magnetic field of a pole strip at an air gap of 1 mm 56
Figure 5.20: Performance function of the adaptive algorithm using physical model 59
Figure 5.21: 3D learning function of the adaptive algorithm using simplex method 60
Figure 5.22: 2D learning function of the adaptive algorithm using simplex method 60
Figure 5.23: Magnetic field at different revolution speeds 62
Figure 5.24: Normal component of the pulse shaped magnetic field 63
Figure 5.25: Pulse shaped magnetic field of a single magnetic region 63
Figure 5.26: Pulse shaped magnetic field of a single magnetic region 64
Figure 5.27: Digital data at different data rate and corresponding analog voltage 65
Figure 5.28: Predictive decision feedback equalizer with PLL 65
Figure 5.29: Digital phase locked loop used by the pDFE 66
Figure 5.30: Bode diagram of the analog loop filter 66
Figure 5.31: Characteristic of the closed phase locked loop 67
Figure 5.32: All-Digital PLL used by the pDFE 67
Figure 5.33: Bode diagram of digital loop filter 68
Figure 5.34: Frequency fluctuations of the pattern frequency 69
Figure 5.35 Simulation results of an ADPLL 70
Figure 5.36: Frequency fluctuations expressed as constant frequency and phase shift 70
Figure 5.37: Predictive all-digital PLL 71
Figure 5.38: Phase prediction methods 71
Figure 5.39: Simulation results of a pADPLL, no speeding up the engine 73
Figure 5.40: Simulation results of a pADPLL, speeding up of the engine 74
Figure 5.41: All-digital PLL with frequency dependent digital loop filter 75
Figure 5.42: Implementation of a variable integer delay line 76
Figure 5.43: Implementation of a variable integer delay line using counters 76
Figure 5.44: a) Four GMR elements in a Wheatstone bridge configuration 78
Figure 5.45: Predictive decision feedback equalizer 79
Figure 5.46: Simulation results of the pDFE on a gap of 3.5 mm: 80
Figure 5.47: Cross section of the performance function 82
Figure 5.48: Calculated differential output voltage of a Wheatstone bridge 83
Figure 5.49: Finite filter length of the pre- and post-FBF 84
Figure 5.50: Three-dimensional performance function 85

List of Acronyms

1/N	frequency divider
ABS	Anti-lock Braking System
AC	Alternating Current
ADC	Analog to Digital Converter
ADPLL	All-Digital Phase Locked Loop
CMR	Colossal MagnetoResistance
CP	Charge Pump
DC	Direct current
DCO	Digitally Controlled Oscillator
DFE	Decision Feedback Equalizer
DLF	Digital Loop Filter
ECU	Engine Control Unit
ESC	Electronic Stability Control
FBF	Feed-Back Filter
FEM	Finite Element Method
FFF	Feed Forward Filter
FIR	Finite Impulse Response
FM	Frequency Meter
G	air Gap between sensor element and pole wheel
GMR	Giant MagnetoResistance
iGMR	integrated GMR
IIR	Infinite Impulse Response
ISI	InterSymbol Interference
LE	Linear Equalizer
LF	Loop Filter
LUT	Look Up Table
M	Magnetic dipole moment
MATLAB	MATrix LABoratory (software)
MR	Magnetic Region
pADPLL	Predictive All-Digital Phase Locked Loop
PAM	Pulse Amplitude Modulation
pDFE	Predictive Decision Feedback Equalizer
PFD	Phase and Frequency Detector
PLL	Phase Locked Loop
PRML	Partial Response Maximum Likelihood
SARADC	Successive Approximation Register ADC
TDC	Time to Digital Converter
TMR	Tunnel(ing) MagnetoResistance
VCO	Voltage Controlled Oscillator
VD	Variable Delay line
VHDL	VHSIC Hardware Description Language
VHSIC	Very-High-Speed Integrated Circuits

CHAPTER 1

Introduction

Over recent decades, the use of electronic sensors in automotive applications has greatly increased. This trend began in 1960, with the installation of an electronic sensing arrangement for engine management. In the '80s, passenger protection became more important, which resulted in additional applications such as air bags and the anti-lock braking system (ABS). Today, every new car worldwide includes an average of approximately 25 sensor chips. This number is still increasing.

Not only the number of sensors, but the functionality of the sensors also increased over the past few decades. In the early stage of electronic measurements, a sensor was used to convert physical values such as the magnetic field, pressure, or temperature into an electronic signal such as voltage or current. The output of the sensor element was evaluated by a microcontroller which calculated the physical value.

High integration levels of today's semiconductor technologies allow complex signal processing to be integrated on a small silicon area. The trend of integrated sensors serves to optimize one sensor and its corresponding complex signal processing to a specific application. Using this approach, systematic errors of the sensing arrangement can be compensated on sensor chip.

The thesis focuses on a sensing arrangement for angle measurement based on magnetic measurements. In this arrangement, a systematic angle error of up to 1.8° appears. Use of digital signal processing should compensate for the error.

Chapter 1 – Introduction

The thesis is organized into the following chapters:

- Chapter 2 gives a short review of decision feedback equalizers, adaptive algorithms, and phase locked loops fundamentals.
- Chapter 3 presents an overview of the measurement setups for angle measurements used in automotive applications. The typically used sensing arrangements based on magnetic measurements are described in more detail.
- An overview of sensor concepts in use today, which do not compensate the systematic angle error, is described in chapter 4. Simulation results show the waveforms of the magnetic field and the systematic error caused by air gap variations.
- Chapter 5 presents one possible solution that compensates for the displacement. The presented solution consists of a modified DFE, an adaptive algorithm, and a modified PLL. Each part is explained separately, and its functionality is illustrated by presenting simulation results. The chapter concludes with simulation results of the digital filter in the time domain and performance limit calculations.
- Chapter 6 concludes the thesis with a short summary.

CHAPTER 2

A Brief Review of Fundamentals

2.1 Decision Feedback Equalizer Fundamentals

The decision feedback equalizer is a filter structure which is commonly used to remove, for instance, intersymbol interference (ISI) caused by a transmission channel. This filter structure consists of two linear equalizers and a nonlinear slicer (discriminator). The linear equalizers remove the ISI and the slicer generates the noiseless binary output signal.

2.1.1 Linear Equalizer

Generally, the linear equalizer (LE) is a filter in the analog or digital domain. A digital LE can be implemented either as a finite impulse response (FIR) or infinite impulse response (IIR) filter [Qureshi'82, Qureshi'85]. Figure 2.1 shows a digital LE operating on the output $r[c]$ of a digital transmission channel with white noise. To remove all ISI of the channel, the transfer characteristic for the so-called »zero forcing linear equalizer« is given by

$$g_r[c] = \frac{q[c]}{g_{sc}[c]}, \qquad (2.1)$$

where $q[c]$ is the wanted shape of the pulse after the LE. In the absence of noise, the reconstructed data $\hat{m}[c]$ is equal to the transmitted data $m[c]$ [Bergmans'96]. However, if white noise appears, the LE can have infinite noise enhancement. For example, on a channel with low pass characteristics the linear equalizer with transfer characteristics (2.1) enhances noise at high frequencies [Bergmans'96, Le'02].

Therefore, the so-called »minimum mean square error linear equalizer«, which minimizes the compound power of ISI and noise, is more commonly used. The transfer characteristic of the LE is calculated by minimizing the power of the error signal [Bergmans'96]

$$e[c] = (\tilde{m}[c] - \hat{m}[c])^2. \tag{2.2}$$

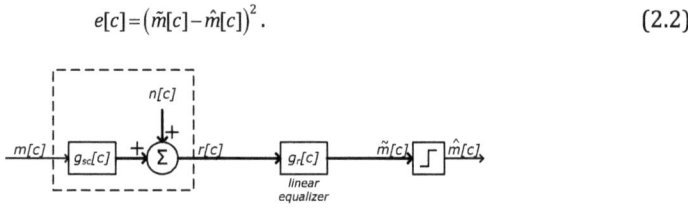

Figure 2.1: Linear equalizer $g_r[c]$ and digital transmission channel $g_{sc}[c]$ with noise $n[c]$

2.1.2 Decision Feedback Equalizer

The linear equalizer can also be realized as a filter with transfer characteristic $g_{sc}[c]-1$ placed in a feedback loop (Figure 2.2.a) [Barry'04]. The transfer characteristics of this feedback system is given by

$$g_r[c] = \frac{1}{1+(g_{sc}[c]-1)} = \frac{1}{g_{sc}[c]} \tag{2.3}$$

and is equal to the characteristic (2.1) of the LE in Figure 2.1.

In the absence of noise, the input and output of the slicer are equal. The slicer has no effect and therefore the characteristics of the LE are not affected by moving the slicer inside the feedback loop (Figure 2.2.b). Moving the slicer inside the feedback loop, the noise enhancement is reduced because the slicer removes the circulating noise within the loop [Bergmans'96, Rothenberg'97].

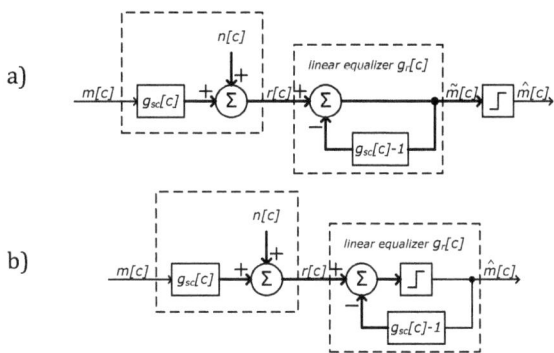

Figure 2.2: Linear equalizer realized as a filter in the feedback path
a) slicer after the LE
b) slicer within the loop

Since the output of the slicer is an estimate $\hat{m}[c]$ of the current datum $m[c]$, the filter in the feedback path can only generate ISI of the last data to the current datum (called post-ISI, Figure 2.3.b) [Barry'04, Brown'99].

Because future data is unknown, the same approach cannot be used to calculate the interference of future data to the current datum (called pre-ISI). An additional filter, called feed forward filter, is required to cancel the pre-ISI [Barry'04, Bergmans'96, Brown'96, Brown'99, Kajley'97]. The resulting filter structure, employing a filter in the feed forward and feedback path to remove pre- and post-ISI, respectively, is called a decision feedback equalizer (DFE, Figure 2.4) [Austin'67].

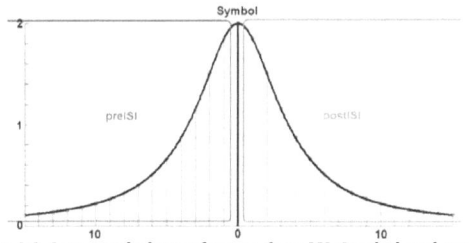

Figure 2.3: Intersymbol interference (pre-ISI, Symbol and post-ISI)

A drawback of the DFE is error propagation. If an incorrect decision was made, the incorrect decoded data remains in the feedback loop and increases the probability of future incorrect decisions. However, this error propagation is reduced by the slicer [Barry'04, Bergmans'96, Qureshi'82].

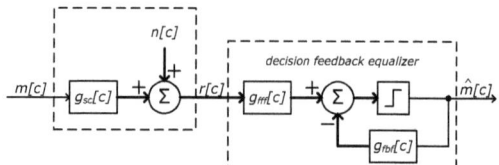

Figure 2.4: Decision feedback equalizer

Most DFEs are realized with FIR filters that are clearly stable [Barry'04, Mc Clellan'99]. Due to their finite impulse response, the output of FIR filters settles to zero. Using the output of FIR filters to remove pre-ISI or post-ISI, the tails of the signal remain prior to discrimination. A remaining small tail is similar to noise and is removed by the slicer. However, for the theoretical considerations in this chapter, an ideal FIR filters with infinite filter order is assumed [Qureshi'82].

Again, the filter coefficients of the DFE can be calculated to minimize the ISI of the channel or the compound power of ISI and noise.

2.1.3 Partial Response Technique and Decision Feedback Equalizer

The partial response technique is used to transform the data sequence $m[c]$ into a symbol sequence $a[c]$. If the frequency spectrum of the symbol sequence is better matched to the channel characteristic than the spectrum of the data sequence, distortion of the channel can be avoided [Bergmans'96].

The mapper in Figure 2.5.a transforms the binary data sequence $m[c] \in \{-1,1\}$ in a ternary symbol sequence $a[c] = m[c] - m[c-1]$ with $a[c] \in \{-2,0,2\}$. Assuming an ideal transmission channel with no ISI, the transmitted data can be reconstructed using the partial response decoder $\tilde{m}[c] = a[c] + \tilde{m}[c-1]$ at the receiver. Knowing the last datum $\tilde{m}[c-1]$, the current datum $\tilde{m}[c]$ can be calculated.

However, for $k=0$ the last datum is unknown and an assumption must be made. An incorrect assumption for $\tilde{m}[0]$ yields the next datum $\tilde{m}[1] \in \{-3,+3\}$ which is a reconstruction error. Error propagation, caused by the feedback path, leads the following data sequence to $\tilde{m}[c] \in \{-3,-1\}$ or $\tilde{m}[c] \in \{+1,+3\}$ [Bergmans'96]. Therefore, a slicer is used to force data to be $\hat{m}[c] \in \{-1,+1\}$ and thus to avoid error propagation (Figure 2.5.b).

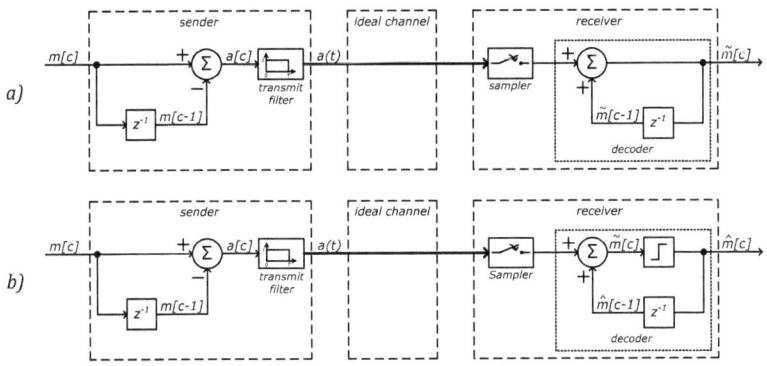

Figure 2.5: Partial response equalizer
a) Sender, ideal channel and receiver
b) Sender, ideal channel and receiver with slicer

In a real transmission channel, distortions appear which result in a broadening of the pulse shape (Figure 2.3). The transmitted symbols can be estimated from $v(t)$ only if ISI has been removed which can be done using a DFE. The partial response decoder and the DFE can be combined as shown in Figure 2.6.

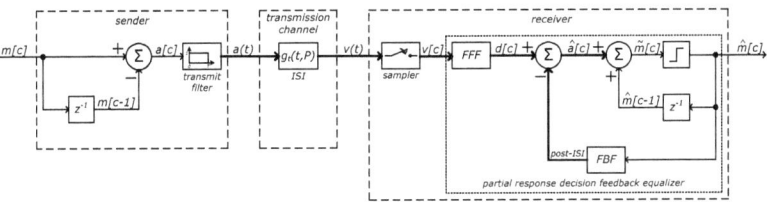

Figure 2.6: Partial response decision feedback equalizer

2.1.4 Adaptive Decision Feedback Equalizer

The decision feedback equalizer (DFE) can remove ISI if the transfer characteristic (distortion) of the channel is known. Usually the distortion is unknown and adaptive algorithms are used to estimate the filter coefficients of the DFE.

In digital communication, training sequences are commonly used to estimate the filter coefficients [Brown'96, Brown'99, Götze'00, Qureshi'82, Treichler'01]. The training sequence $l[c]$—which is known at the receiver—is transmitted through the unknown transmission channel. The adaptive algorithm at the receiver compares the signal $\tilde{m}[c]$ (the signal after partly removing ISI) with the known training sequence $l[c]$. The

CHAPTER 2 – A BRIEF REVIEW OF FUNDAMENTALS

squared difference between these two signals, called error signal [Brown'99, Kajley'97]

$$e[c] = (l[c] - \tilde{m}[c])^2, \qquad (2.4)$$

identifies remaining ISI in $\tilde{m}[c]$ and is used to estimate the filter coefficients. Minimizing the error signal, the filter coefficients converge to their optimal values. If the error signal has reached its smallest possible value, the optimal filter coefficients to remove ISI are found [Cioffi'90, Haykin'01, Qureshi'82].

After adaption, the data transmission starts and data is transmitted instead of the training sequence. Using the squared difference between the signal $\tilde{m}[c]$ and the decoded data $\hat{m}[c]$ as a new error signal

$$e[c] = (\hat{m}[c] - \tilde{m}[c])^2 \qquad (2.5)$$

(see Figure 2.7), the estimation of the filter coefficients can continue.

In the presence of noise, the compound power of ISI and noise is minimized if (2.4) and (2.5) are used as error signals.

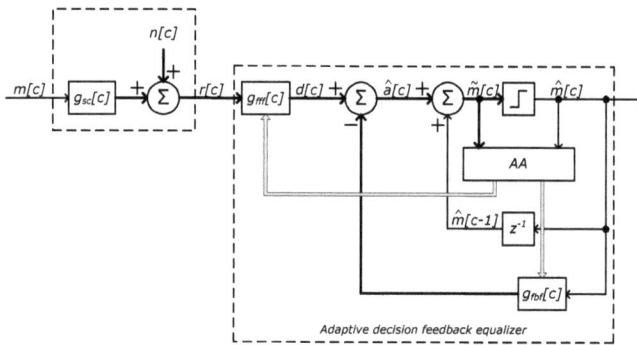

Figure 2.7: Adaptive decision feedback equalizer

2.2 Adaptive Algorithm Fundamentals

To illustrate the fundamentals of filter coefficient estimation using adaptive algorithms, a simple example is presented [Treichler'01]. In this example an unknown integer n_u between 0 and 100 is estimated by an adaptive algorithm. Beginning with a first guess $n_g[1]$ for the number, the next guess should be closer to the unknown number. To gauge the quality of the guess, an error function e[i] is defined. In this example, an error function is chosen which can perform following comparisons: the guess was higher, lower or equal to the unknown number. This ternary error signal can be mathematically written as

$$e_t[i] = \text{sgn}(n_u - n_g[i]) \tag{2.6}$$

where

$$\text{sgn}(x) = \begin{cases} +1 & \text{if } x > 0 \\ 0 & \text{if } x = 0 \\ -1 & \text{if } x < 0 \end{cases} \tag{2.7}$$

If the guess was too high ($e_t[k] = -1$) or too low ($e_t[k] = +1$), the next guess should be a smaller or lower number, respectively [Treichler'01]. This way, the value of the guess converges to the unknown number n_u:

$$\lim_{i \to \infty} n_g[i] = n_u \tag{2.8}$$

2.2.1 Performance Function

The so-called performance function shows the value of the error signal as a function of all adaptive algorithm parameters [Treichler'01]. Figure 2.8.a shows the performance function of the number guessing problem using the ternary error signal (2.6). The performance function has a zero if the guessed number is equal to this unknown number, which is 40 in this example.

The same problem can also be solved by using the squared error signal

$$e_{sq}[i] = (n_u - n_g[i])^2, \tag{2.9}$$

and the corresponding performance function is shown in Figure 2.8.b. The squared error signal increases with an increasing difference between the unknown and the guessed number. The global minimum of the performance function can be found by minimizing the error signal. If local minima are present in the performance function, the localization of the global minimum is more complicated [Ramsden'06, Widrow'85]. The adaptive algorithm could stuck in a local minimum instead of finding the global one.

CHAPTER 2 – A BRIEF REVIEW OF FUNDAMENTALS

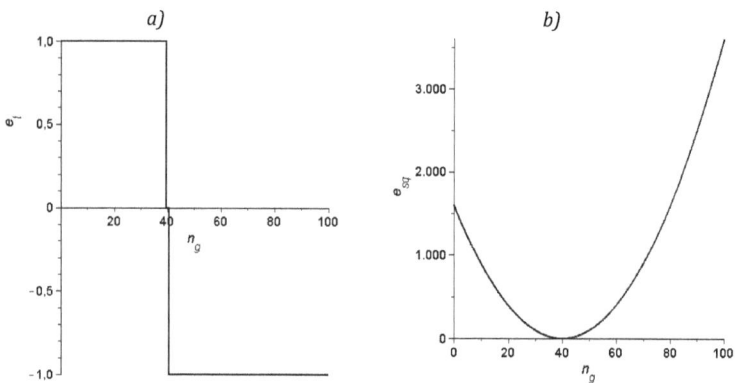

Figure 2.8: Performance function of the number guessing problem
a) ternary error signal
b) squared error signal

2.2.2 Adaptive Methods

Many different adaptive methods can be used to find the global minimum of a performance function i.e. to find the unknown parameter value. Some of them use the ternary error signal (2.6), but most of them use the squared error signal (2.9) which allows estimating the unknown number using fewer iteration steps.

The simplest adaptive method uses the ternary error signal and varies the guessed number by a constant step size μ (e.g., 1) until the error signal is zero, i.e.

$$n_g[i+1] = n_g[i] + \mu \cdot e_t[i]. \tag{2.10}$$

A more advanced method, using the ternary error signal and a variable step size, is the so-called bisection method. The first guess is in the middle of the interval. Evaluating the error signal, the subinterval that includes the unknown number can be determined. The next guess is in the middle of the interval which includes the unknown number. By repeating this process, the guess converges to the unknown number [Burden'00].

The gradient method evaluates the squared error signal (2.9). The step size depends on the gradient of the performance function at the position of the previous guess [Treichler'01]:

$$n_g[i+1]=n_g[i]-\mu\cdot e'_{sq}[i]=n_g[i]-\mu\cdot\frac{\partial e_{sq}[i]}{\partial n_g[i]} \qquad (2.11)$$

Approximating to the unknown number, the gradient of the performance function decreases and consequently the step decreases as well (Figure 2.9.a). With the decreasing step size, the unknown number is estimated more accurately with each iteration step.

Newton's method, an algorithm for finding the root (zero point) of a function, calculates the next guessed number by

$$n_g[i+1]=n_g[i]-\frac{e_{sq}[i]}{e'_{sq}[i]}=n_g[i]-\frac{e_{sq}[i]}{\frac{\partial e_{sq}[i]}{\partial n_g[i]}}. \qquad (2.12)$$

This method finds zeros—and not minima—in the performance function. If the error signal at the position of the global minimum is not zero, the position of the minimum can be found by searching a zero in the first derivative of the function:

$$n_g[i+1]=n_g[i]-\frac{e'_{sq}[i]}{e''_{sq}[i]}. \qquad (2.13)$$

Higher order derivatives of the performance function are also used by Halley's and Householder's method [Deuflhard'04]. These methods are based on Newton's method and search for zeros in the performance function.

Chapter 2 – A Brief Review of Fundamentals

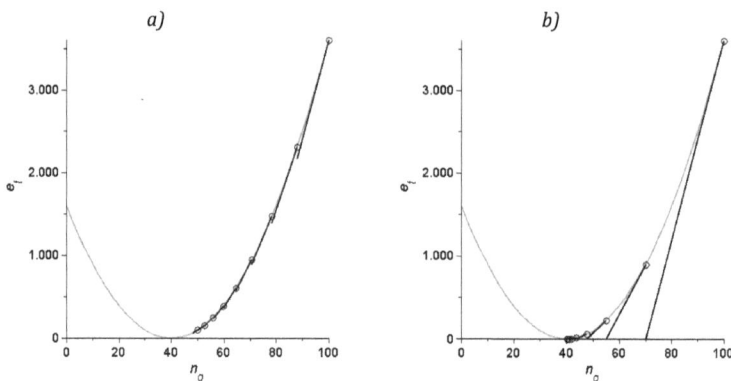

Figure 2.9: Illustration of adaptive methods
a) Gradient method with $\mu = 0.1$
b) Newton's method

The Levenberg-Marquardt [Levenberg'44] method assumes a performance function with parabolic shape. A two-dimensional performance function (one parameter and the error signal) with parabolic shape can be described by three points on the function. Knowing three points on a parabola, the parabolic equation can be calculated. Using this approach, three points on the performance function are determined and the parabolic equation is calculated. The position of the parabola minimum is the next guessed number.

A different approach, calculating the next guess without using extrapolation or fitting, is the so-called simplex method. Using this method, the next guessed number is calculated by comparing multiple values of the error signal at different positions on the performance function. To illustrate the principle of this method, a three-dimensional performance function, having two unknown numbers n_1 and n_2, is assumed. A 2-simplex (i.e. triangle) is defined in the parameter plane (Figure 2.10.a) and the error signal is determined on each end-point of this simplex. The end-point with the largest error signal is reflected through the centroid of the remaining 1-simplex (i.e. line). A new simplex is defined by replacing the end-point with the largest error signal with the point obtained by reflection (Figure 2.10.b). In the following iteration steps, the value of the error signal at the position of the new end-point of the simplex is required.

An enhanced simplex method that varies the size of the simplex after reflection, was presented by Nelder and Mead in [Nelder'64]. If the value of the error signal at the new position is smaller than all values of the old simplex, the new simplex is expanded in the direction of the new end-point (Figure 2.10.c). If the value at the new end-point is higher, the simplex is shrinked [Hudzovic'01].

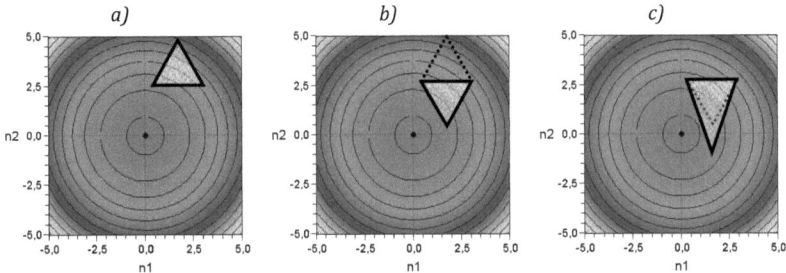

Figure 2.10: Illustration of the simplex method
a) simplex with three end points
b) new simplex obtained by reflection of the end-point with the largest error signal
e) expansion of a simplex (Nelder-Mead method)

2.2.3 Learning Function

The learning function of an adaptive algorithm shows the calculated parameter value as a function of the number of iteration steps. This function shows graphically the adapting speed and steady state accuracy of an adaptive method.

The learning functions of the number guessing problem using constant step size is shown in Figure 2.11.a. The value of the guess approximates the unknown number. In the steady state, the guessed number is alternately higher and lower than the unknown number. Decreasing the step size causes a wanted increase in steady state accuracy but an unwanted decrease in adapting speed. A tradeoff between speed and accuracy must be found.

Better performance can be achieved if a variable step size is used. For example, the step size can be decreased if an oscillation is detected (similar to Nelder-Mead) which increases the steady state accuracy (Figure 2.11.a). Using the gradient or Newton's method, the step size decreases with the gradient of the performance function and the value of the guess converges to the unknown number (Figure 2.11.b and .c).

Compared to complex algorithms, simple algorithms need more iteration steps to estimate the unknown number with the same accuracy. On the other hand, complex

CHAPTER 2 – A BRIEF REVIEW OF FUNDAMENTALS

algorithms require the gradient of the performance function; determining the gradient of a function is typically more complex than determining the value of a function.

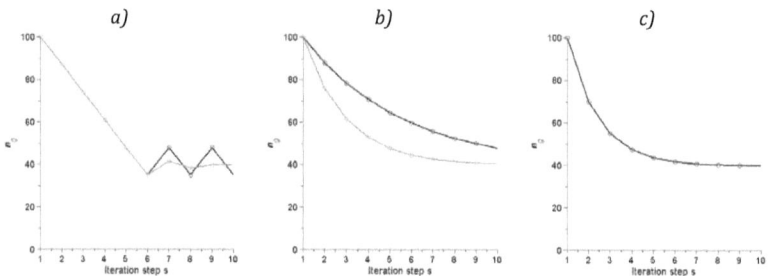

Figure 2.11: Learning function using different adaptive methods
a) dark blue: constant step size of 13; light blue: decreasing step size during oscillation
b) gradient method; dark blue: $\mu = 0.1$; light blue: $\mu = 0.2$
c) Newton's method

In a typical application, the value of the error signal is obtained by measurements. If the gradient of the error signal cannot be measured, it can be approximated by the finite difference

$$\frac{\partial e_{sq}[i]}{\partial n_g[i]} \cong \frac{\Delta e_{sq}[i]}{\Delta n_g[i]}. \qquad (2.14)$$

With this approximation, two measurements on each iteration step are necessary and a distinction must be made between the number of iteration steps and the number of measurements. (Second and higher order derivatives can also be approximated by finite differences and three or more measurements on each iteration step are necessary.)

2.2.4 Adapting Multiple Parameters

If an adaptive algorithm has to estimate more than one unknown number, the adaptive method has to find the global minimum of a multi-dimensional performance function. The adaptive methods summarized above can be used if the scalar parameter n_g is replaced by the parameter matrix $\mathbf{n}_g$. (The error signal remains a scalar.) More details about adapting multiple parameters can be found in [Bergmans'96, Haykin'01, Treichler'01] and the above referenced literature.

2.3 Phase Locked Loop Fundamentals

2.3.1 Digital Phase Locked Loop

Figure 2.12 shows the block diagram of the so-called »digital phase locked loop« which consists of a digital phase and frequency detector (PFD), an analog charge pump (CP), an analog loop filter (LF) an analog voltage controlled oscillator (VCO) and a digital frequency divider (/N). Each sub-block of the phase locked loop (PLL) can be realized in different ways [Best'04] resulting in a variety of digital PLLs [Gupta'75]. The digital charge pump PLL is commonly used [Da Dalt'07] and therefore described in this section.

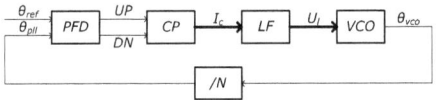

Figure 2.12: Digital charge pump phase locked loop

A simple PFD consists of two flip flops and an AND-gate and converts the phase difference between the binary signals θ_{ref} and θ_{pll} into the digital control signals UP and DN (Figure 2.13.a).

The CP transforms the three stable states of the PFD into the current $I_c \in \{+1, 0, -1\}$. This current is pulse-width modulated and consists of both, DC and AC components [Best'04]. If the phase difference between the binary signals θ_{ref} and θ_{pll} is in the phase range from -2π to 2π, the DC component is proportional to the phase difference (Figure 2.13.b) whereas the AC component is unwanted noise.

The LF suppresses the AC component in the CPs output current and delivers a DC voltage for the VCO. The dynamic behavior of the PLL (rise time, frequency response) can be defined by designing the LF.

The resonance frequency of the VCO can be varied by the output voltage of the loop filter by using varicap diodes (diodes with a voltage controlled capacity) in the resonator of the VCO. In the phase domain (Figure 2.12), the VCO has integrating behavior and its transfer characteristic is given by K_0/s [Best'04].

CHAPTER 2 – A BRIEF REVIEW OF FUNDAMENTALS

In the locked state, the signals θ_{ref} and θ_{pll} have the same phase and frequency. Using the frequency divider /N the oscillation frequency of the VCO is N times higher than the frequency of the reference signal θ_{ref}.

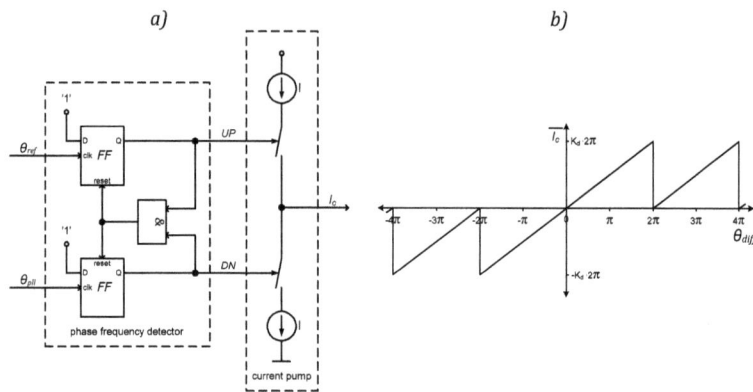

Figure 2.13: Phase and frequency detector and current pump
a) block diagram
b) averaged output current $\overline{I_c}$ versus phase difference θ_{diff} between θ_{ref} and θ_{pll}

2.3.2 All-Digital Phase Locked Loop

Figure 2.14 shows the so-called »all-digital PLL« where each analog block of a »digital PLL« (Figure 2.12) is replaced by its digital equivalent. The current pump is replaced by a time to digital converter (TDC), the analog loop filter is replaced by a digital loop filter (DLF) and the VCO is replaced by a digitally controlled oscillator (DCO). In the new structure, multi-bit signals are used instead of analog signals resulting in discretization and quantization effects. Quantization effects are discussed in more detail in [Madoglio'07].

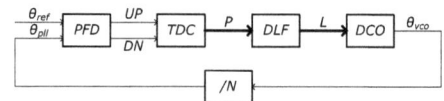

Figure 2.14: All-digital phase locked loop

The TDC converts the length of the UP and DN pulses into the digital number P. Usually a counter with constant sampling time is used to measure the pulse length resulting in a discrete phase resolution (Figure 2.15) [Kratyuk'07]. If the pulse length

CHAPTER 2 – A BRIEF REVIEW OF FUNDAMENTALS

of the UP and DN pulses is smaller than the sampling time of the TDC, its output is zero. Effects of this dead band (i.e. hysteresis) are discussed in [Lee'04]. If the length of the UP and DN pulses is much longer than the sampling time of the TDC the quantization errors can be neglected [Kratyuk'07].

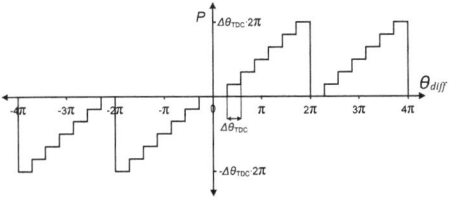

Figure 2.15: Transfer characteristic of the PFD followed by the TDC the sampling time T_{tdc} of the TDC is expressed as phase resolution $\Delta\theta_{TDC} = 2\pi \cdot T_{tdc}/T_{ref}$ [Kratyuk'07]

The digital equivalent of the analog loop filter is the digital loop filter. Its parameters can be obtained by transforming the continuous time transfer function into discrete time domain. Discretization effects cause differences between the frequency response of the analog and digital loop filter. These differences can be minimized by using a high sampling frequency.

The digital loop filter is mostly an IIR filter which can be implemented in the canonical or biquad form. Theoretically, the frequency responses of both filter forms are equal, but quantization errors influence their behavior. Canonical IIR filters are more vulnerable to quantization errors than biquad ones [Hammerschmidt'07]. Therefore, the biquad filter (Figure 2.16) is recommended for implementation.

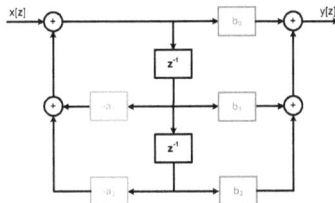

Figure 2.16: Second order biquad IIR filter with transfer function (2.15)

$$h[z] = \frac{y[z]}{x[z]} = \frac{b_0 + b_1 \cdot z^{-1} + b_2 \cdot z^{-2}}{1 + a_1 \cdot z^{-1} + a_2 \cdot z^{-2}} \qquad (2.15)$$

A simple solution of a DCO uses a divide-by-N counter to scale down a high frequency clock signal [Best'04] by a variable scale down factor. The output frequency f_k of the DCO is quantized to

$$f_k = \frac{1}{N} \cdot f_{high} \tag{2.16}$$

where N is the integer scale down factor.

In contrast to a VCO, the output frequency of the DCO is inversely proportional to its control input N. Therefore, a L to N converter, with $N=1/L$, is used inside the DCO to make its output frequency directly proportional to its input L.

CHAPTER 3

Speed and Angle Measurements in Automotive Applications

3.1 Measurement Techniques

Speed and angle measurements of rotating shafts are very important in the area of mechanical engineering; they have 20% of the gross sales revenue of all automotive sensors [Wolber'77]. Typical applications in the automotive area are speed and angle measurements of the engine crankshaft (required by engine control unit (ECU)), speed measurements of the wheels (required by anti-lock braking system (ABS)), angle measurements of the steering wheel (required by drive by wire, electronic stability control (ESC)), etc.

Engine applications are challenging due to their requirements on high angle accuracy under harsh environmental conditions such as temperature variations between −40°C and +125°C, humidity variations between 0% and 100% and contamination (oil, water, salt, dust, etc.).

In general, there is a wide variety of speed and angle measurement techniques, but only a few of them are used in engine applications. In the following sections, the most commonly used techniques are briefly described. Other techniques can be found in [Lequesne'98, Lequesne'99, Marks'80, Wolber'77].

3.1.1 Potentiometers

Potentiometers were the first used angle sensors in the automotive area. They were introduced in the middle eighties and are still used in a number of applications (e.g., throttle valve, fuel tank, etc.) [Hobein'04]. Potentiometers combine high accuracy with very low cost. The disadvantages of potentiometers are their abrasive behavior resulting in a short lifetime and their vulnerability to high temperature and contamination [Hella'03]. Potentiometers have a limited range of safety features, which is a disadvantage for X-by-wire[1] applications [Granig'07b].

3.1.2 Rotating Permanent Magnet

A permanent magnet is mounted at the end of the shaft. If the shaft rotates, the magnetic field around the magnet rotates with the shaft. A magnetic field sensor element—mounted on the rotation axis of the shaft—measures the direction of the field. The output of the sensor element is the cosine of the angle between magnet and sensor. Two sensor elements, placed at 90° angle to each other, are used and they deliver the sine and cosine of the angle. Using the arctangent function, the angle can be calculated from the sensor signals. More details about this arrangement are given in [Granig'07c, Hammerschmidt'05].

Mounting the sensor on the rotation axis of the shaft limits the number of possible applications. This arrangement is commonly used to measure the angle of the steering wheel. The resolution can be below 0.03° with an accuracy of about 1.4° [Granig'07b].

3.1.3 Resolver

A resolver consists of a rotor coil that is mounted on the rotating shaft and two stator coils that are placed at 90° angle to each other. A current with a sinusoidal shape and a constant peak value is induced in the rotor coil. The electromagnetic field of the current flowing through the rotor coil rotates with the shaft. The rotating field induces a current into the two stator coils. These currents have a constant frequency but their amplitude is angle-dependent (i.e. the sine and cosine of the shaft angle). The angle can

---

[1] *The automotive trend to replace mechanical components with electrical sensors and actuators is called X-by-wire.*

be calculated by determining the peak values of the sinusoidal currents and using the arctangent function. More details are presented in [Ellis'01, Granig'07b].

Resolvers are very robust to contamination and mechanical vibrations. Their angle accuracy is high and can be less than 0.1° [Granig'07a]. This makes them predisposed for the automotive market; however, the disadvantages of resolvers are their demand of additional electronics for signal generation and signal processing, their large size and their very high costs [Granig'07a].

3.1.4 Rotary Encoder

A rotary encoder consists of a patterned wheel that is mounted on the rotating shaft and a sensor element. The pattern on the wheel can be conductive, optical, capacitive, inductive, magnetic, etc. If the wheel rotates, the pattern on the wheel passes the sensor element and the sensor generates a signal which is an electric image of the pattern on the wheel. Knowing the pattern, the rotational speed and angle information can be determined from the sensor signal.

The use of a wheel with multiple tracks and multiple sensor elements is also common. Using multiple tracks, the sensor elements deliver a multi-bit signal and the angle can be determined also if the wheel does not rotate.

The patterned wheel of a conductive rotary encoder consists of conductive and nonconductive regions. The rotation of the wheel connects and disconnects two contacts and behaves like an electrical switch. This technique is very simple, but it has many disadvantages such as abrasive behavior, wiper bounce, limited angle accuracy and vulnerability to contamination.

Optical rotary encoders comprise a light source, a code disk and an optical sensor element. They are well established in a broad area of different applications and enable highest accuracies down to 300×10^{-9} ° [Kojima'04]. However, high resolution requires a fine structure on the code disk, which is expensive in production and vulnerable to contamination. Therefore, only rotary encoders with lower resolution are used in the automotive area [Hobein'04]. A typical application is the angle measurement of the steering wheel where a resolution of 1.5° to 3.0° is sufficient [Granig'07a]. Optical rotary encoders were also used in crankshaft applications but they have limited resistance to contamination and high temperature. The latter problem can be avoided by using fiber optics, which results in higher costs [Wolber'77].

Rotary encoders using the capacitive method consist of a segmented sending plate, a receiving plate and a rotor. The angular movement of the rotor changes the capacity between the electrode segments on the sending and receiving plate [Hu'03, Sauter'02, Sauter'05]. This principle is well established in science and industry [Zangl'02] but not (yet) very commonly used in the automotive area. See [Hu'03, Sauter'02, Sauter'05] for more details.

The inductive sensing arrangement consists of a transmission coil, one or more receiving coils and a rotor. An AC current flows through the transmission coil and generates an electromagnetic field which is influenced by the rotor. The influenced electromagnetic field induces an angle-dependent electrical voltage in the receiving coils. The inductive principle is robust to dirty environment, high temperature and mechanical vibrations [Hella'03]. A disadvantage of the inductive principle is that the operating range is limited to 120°. Measurements over a full rotation are only possible by using subunits which leads to points of discontinuity [Granig'07b]. The inductive principle is not well established in the automotive area because the complex mechanical structures result in high costs. However, there is a solution to this problem (presented in 2004 in [Hobein'04]).

The sensing arrangement based on magnetic measurements comprises a magnetically patterned wheel and a magnetic field sensor element. This arrangement is described in more detail in Section 3.3. The magnetic principle is very robust to dirty environment, contamination, temperature variations and allows contactless measurements [Zangl'02]. These advantages and the low production cost of magnetic field sensor elements make the magnetic principle to the favorite principle in the automotive area in the recent decades. Magnetic field sensors have the largest market share of all sensor types [Marek'03] and are commonly used for a wide range of different applications such as camshaft, crankshaft, ABS, ESC, wind shield wiper, etc. The main disadvantage of sensing arrangements based on magnetic measurements is their low angle accuracy, which is in the range between 1° to 2°.

The angle resolution of a rotary encoder primarily depends on the geometrical length of the coded regions on the wheel. However, on capacitive, inductive and magnetic coded wheels, a smooth transition between the states of the sensor output signal can be observed. Applying signal processing to the analog sensor signal allows an increase in the resolution far below the geometrical length of the coded regions. More details about this signal processing are given in [Sauter'05] for the capacitive principle, [Hobein'04] for the inductive principle and in this thesis for the magnetic principle.

3.2 Engine Application

Most cars in use today employ four stroke combustion engines to convert the energy stored in fuel into a mechanical force. The maximal rotational speed and idle speed of an automotive diesel engine crankshaft vary between customers and are in the range of 6,000 to 10,000 rpm and a few hundred rpm, respectively. The rotational speed of the camshaft is half of the speed of the crankshaft.

During combustion stroke, the rotational speed of the engine crankshaft is accelerated by the impulsive force of the piston which results from the explosion of the fuel in the cylinder. During intake, compression and exhaust stroke, the engine piston requires energy, resulting in a decreasing angular speed[2]. Figure 3.1.a shows the typical speed fluctuations of a four cylinder four stroke diesel engine crankshaft at an averaged rotational speed of 1090 rpm (i.e. 114.1 rad/s). Each cylinder of the engine has one combustion stroke per two rotations. In the ideal case, the forces of each cylinder are equal to each other and the speed fluctuations are periodic if the average rotational speed is constant.

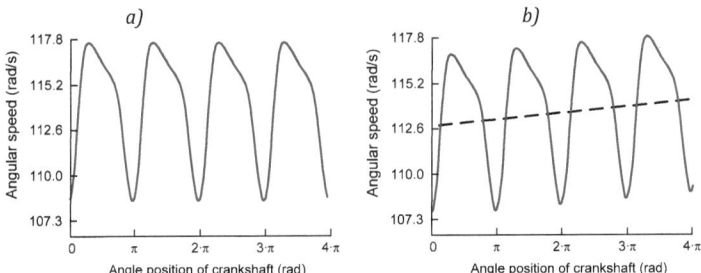

Figure 3.1: Angular speed of engine crankshaft at
a) constant averaged rotational speed [UoM'04]
b) increasing averaged rotational speed with an acceleration of 150 rpm/s (15.7 rad / sec^2)

---

[2] *The angular speed (rad/s) is equivalent to the rotational speed (rev/min) but with different units. Especially in the area of engineering the term rotational speed is more commonly used. However, in this thesis the term rotational speed (rev/min) refers to the average speed over one engine cycle and the term angular speed (rad/s) refers to the speed at the actual angular position.*

The speed variations during acceleration of the engine are small compared to the speed fluctuations (Figure 3.1.b). Only if the engine is accelerated without a load, the speed variations (e.g., 3,000 rpm/s) are in the same range as the speed fluctuations.

3.2.1 Misfire Detection

Evaluating the crankshaft speed fluctuations is a commonly used technique to detect misfire events of a combustion engine. If a misfire occurs, the missing explosion in the cylinder results in a missing torque and consequently a small decrease in the crankshaft angular speed [Jurgen'99]. Many methods in the time or frequency domain are known for misfire detection by monitoring the crankshaft speed [Montani'08, Schmidt'00]. Having an accurate speed information, the reason for the lack of torque such as a bad fuel mixture, lack of compression, lack of spark, wrong spark timing, etc. can be identified [Alkhateeb'02].

Engine misfire events can also be detected by optical combustion sensing or exhaust gas pressure analysis. However, evaluating the speed fluctuations is relatively simple, cheap and effective and is therefore the preferred misfire detection method [Alkhateeb'02].

3.2.2 Camless Engine

The crankshaft of an engine is connected to its camshaft by a belt, chain or cogwheel. Energy of the crankshaft is required to overcome friction and to rotate the camshaft, which opens and closes the valves [AutoWeek'06, Valeo'08]. This energy consumption of the engine causes a decrease in the low-end torque.

Energy consumption can be reduced if the mechanical camshaft is replaced by electronic sensors that measure the angle of the crankshaft and electronic actuators which open and close the valves. [Valeo'08] shows that removing the camshaft allows an improvement in efficiency by 15–20%, an increase in torque by 15–20% and a reduction in emissions by 15–20%. A further benefit of the so-called »camless engines« is their very flexible valve timing. Each valve can be opened and closed separately and at different angle positions of the crankshaft. Varying the valve timing with the rotational speed allows an increase in the torque by up to 100% [Hillier'04]. Deactivating cylinders increases the engine performance at low speeds [Valeo'08]. Additional benefits of the camless engine are its reduced weight and its lower idle speed [Valeo'08]. The problem of using camless engines in mass production is that very accurate, robust and cheap angle sensors are not available today [Valeo'08].

3.3 Sensing Arrangements for Angle Measurement in Engine Application

In general, there are two widely used arrangements for sensing the angle of the crankshaft based on magnetic measurement. In both cases, the arrangement comprises a patterned wheel mounted on a rotating shaft and a magnetic field sensor.

In both sensing arrangements, the magnetic field sensor element converts the applied magnetic field into a (upon first consideration) linearly proportional electrical signal [Hammerschmidt'05]. Signal processing is used to convert the analog waveform into a binary sequence which is an electric image of the pattern on the wheel. Knowing the pattern, the rotational speed and angle of the shaft can be determined from this binary signal.

3.3.1 Pole Wheel

In the first arrangement, the patterned wheel is magnetically coded along its circumference (Figure 3.2). If the pole wheel rotates, the alternating magnetic regions pass the sensor and induce a magnetic field. The normal and tangential components of this field have a sinusoidal-like shape as shown in (Figure 3.3[3]) [Hainz'08a, Hainz'07a]. The amplitude of this waveform depends on the gap between sensor element and pole wheel and the magnetic dipole moments of the magnetic regions on the wheel.

---

[3] *In this thesis, the angle of the wheel is expressed as a geometric length x measured along the circumference of the wheel. The measuring unit is the length of one magnetic region (MR), which is typically 2.5 mm (1 MR = 2.5 mm). On a wheel with 120 magnetic regions, the length of 1 MR corresponds to an angle of 3°.*

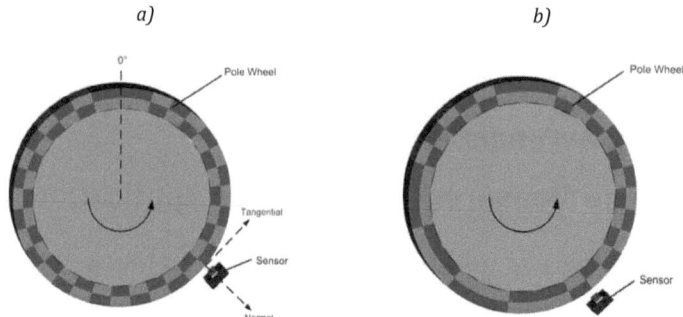

*Figure 3.2: Sensing arrangement for angle measurement using pole wheel
a) patterned pole wheel with one synchronization point b) patterned pole wheel with multiple synchronization points*

Pole wheels with different magnetic patterns are used for a variety of applications. The speed of a shaft (e.g., ABS) can be obtained by using a regular patterned wheel. However, for angle measurements (e.g., crankshaft), an irregular pattern is required. A typical wheel with an irregular pattern is shown in Figure 3.2.a. In addition, wheels with more irregular patterns as shown in Figure 3.2.b are commonly used because they allow synchronization on multiple positions of the wheel. Additional examples of patterned wheels are shown in [Wang'07].

The pole wheel can be manufactured by mounting multiple bar magnets on a carrier wheel. Especially in the early stage of engine management, such wheels were commonly used. However, the recommended small mounting tolerances result in high costs and therefore such wheels are not commonly used today. A cheaper manufacturing process is to magnetize the carrier wheel by a strong magnetic field of a solenoid. Using this process, wheels with different patterns can be manufactured without high effort.

Aging causes a creeping demagnetization of the pole wheel. Temperature variations in the engine compartment result in variations of the wheel magnetic moment. The amplitude of the sinusoidal-like shape of the magnetic field decreases due to aging and varies with temperature.

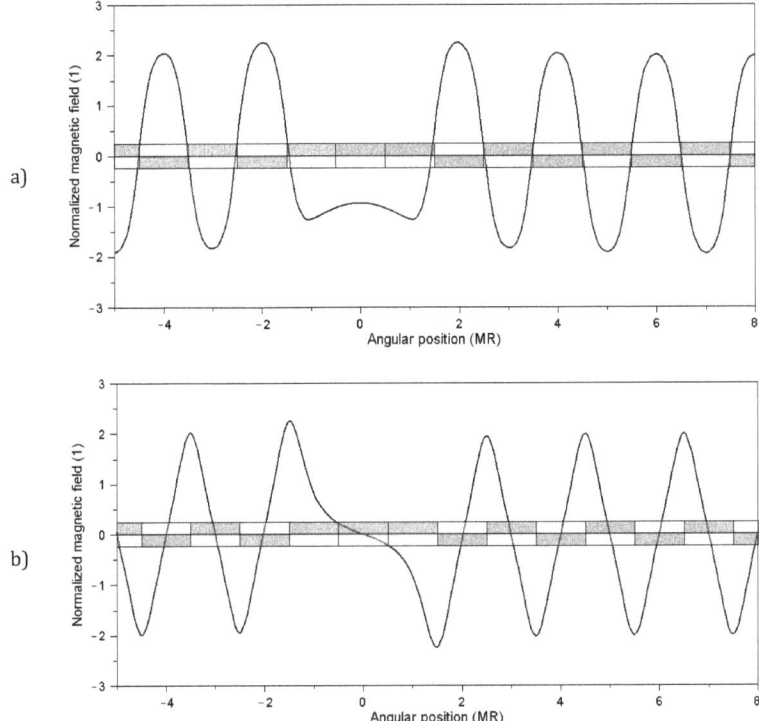

Figure 3.3: Magnetic field of a pole wheel (FEM simulations)
at an air gap of 1.0 mm
a) Normal component
b) Tangential component
[Scherhäufl'06]

3.3.2 Toothed Wheel

The second arrangement uses a toothed wheel which is coded by the length of the teeth and tooth spaces (Figure 3.4.a). A permanent magnet, called »back-bias magnet«, placed behind the sensor creates a constant magnetic field that is influenced by the rotating wheel [Hainz'07a]. If a tooth is in front of the sensor, the field at the sensor position is high and if a tooth space is in front of the sensor, the field is low. Iron casing of the magnet (Figure 3.4.b) can be used to increase the field change at the position of the sensor element.

Chapter 3 – Speed and Angle Measurements in Automotive Applications

The rotating toothed wheel shapes the magnetic field at the position of the sensor element (Figure 3.5.a). The resulting field has the similar sinusoidal-like shape as seen from pole wheels. However, an additional DC component—which is caused by the unipolar nature of the magnetic field—appears and therefore the signal has no zero value (Figure 3.5.a) [Hainz'07a]. Both, the amount of the DC component and the amplitude of the sinusoidal-like waveform, decrease with the gap.

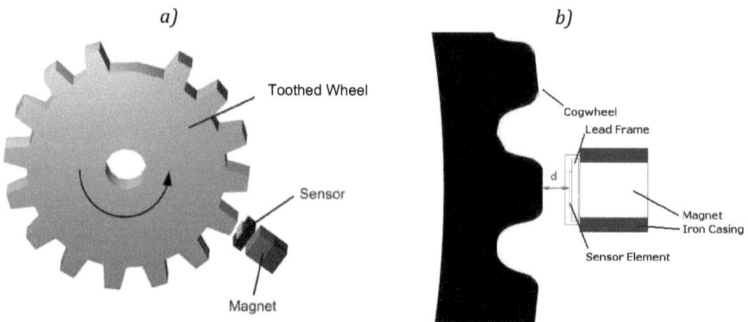

Figure 3.4: Sensing arrangement for angle measurement using toothed wheel
a)patterned toothed wheel with one synchronization point
a) toothed wheel and back-bias magnet with iron casing used in FEM simulations
[Hainz'07a]

The variations of the magnetic field induce eddy currents in the toothed wheel and in the lead frame of the sensor chip. These eddy currents generate their own magnetic field which acts against the original field (Lenz's law). The amount of eddy current depends on the rate of flux variation. This means that at higher rotational speeds, larger eddy currents are generated. Therefore, the total flux density decreases with the rotational speed as shown in Figure 3.5.b [Hainz'07a]. Due to the high electrical conductivity of the toothed wheel and lead frame material, the eddy currents decay rather slowly, which results in a variation of the signal shape [Hainz'07a].

Similar to pole wheels, the magnetic moment of the back-bias magnet decreases due to aging and varies with the temperature.

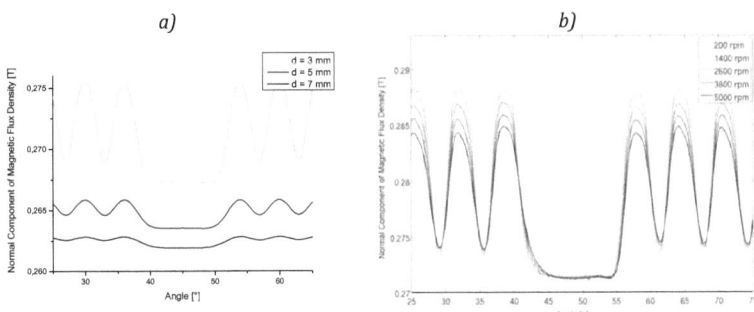

Figure 3.5: Magnetic field of a back-bias magnet, influenced by a toothed wheel (FEM)
a) varying distance between sensor element and back-bias magnet
b) constant distance and varying rotational speed
[Hainz'07a]

3.3.3 Magnetic Field Sensor Element

In both sensing arrangements, magnetic field sensor elements are used to convert the magnetic field into an electrical signal. In general, there exist magnetic field sensors based on a variety of different effects such as the variable reluctance [Wolber'77], the Wiegand effect [Marks'80], the magnetic field sensitive transistor [Halbo'80], etc. All of these sensors can be used to measure the magnetic field; however, in this section only the Hall and magnetoresistance effect are summarized.

The Hall effect was discovered in 1879 by E. H. Hall. If a conductor—through which a current is flowing—is surrounded by a magnetic field a transverse voltage can be measured on the conductor. This voltage is linear in proportion to the normal (Figure 3.2.a) component of the applied magnetic field. Sensors based on the Hall effect can be manufactured in bipolar semiconductor technologies, which allows the integration of the sensor element and signal processing on the same die. In the semiconductor area, sensors based on the Hall effect are most commonly used [Granig'07a]. More details about integrated Hall effect sensors are given in [Ramsden'06].

The magnetoresistance effect was discovered by William Thomson in 1856 when he observed that some materials change their resistance if a magnetic field is applied to them. The observed resistance variations were in the range of a few percent and, therefore, sensors based on this effect are not commonly used. However, this discovery was the basis for the later developed giant magnetoresistance (GMR), colossal magnetoresistance (CMR) and tunnel magnetoresistance effects (TMR).

Some materials showing the GMR effect were discovered in parallel by Peter Grünberg and Albert Fert in 1988; they received the 2007 Nobel Prize in Physics. This effect is still under development, but the manufacturing of sensors based on the GMR effect in CMOS technology is already possible [Hammerschmidt'05]. However, today, Infineon Technologies is the world's only semiconductor supplier to have begun volume manufacturing of integrated sensors based on the GMR effect. These elements are placed on the top of the chip which saves area. The resistance of integrated GMR (iGMR) elements is linearly proportional to the tangential (Figure 3.2.a) component of the applied magnetic field until saturation is reached. Compared to the Hall element, an iGMR element has a higher signal-to-noise ratio [Fleming'01]. More details about iGMR effect sensors are given in [Granig'07b, Granig'07c, Hammerschmidt'05].

CHAPTER 4

Untreated Problem

4.1 Sensor Concepts in Use Today

In the sensing arrangement with a pole wheel, the output voltage of the sensor element is a sinusoidal-like AC waveform. In the sensing arrangement with a toothed wheel, the output voltage is similar but an additional gap-dependent DC component appears. Most sensor concepts in use today estimate and remove the DC component from the sensor voltage. After removing the DC component, the waveform obtained from measurements either using pole wheels or toothed wheels are similar. This allows applying the same subsequent angle and speed measurement techniques to both arrangements. Therefore, removing the DC component is also common on pole wheel measurements.

Different analog or digital solutions can be used to estimate the DC component by evaluating the sensor voltage $v(x)$ (Figure 4.1.a) or the binary sequence $\hat{m}[k]$ (Figure 4.1.b).

The complexity of angle and speed measurement can be reduced by converting the sinusoidal-like AC waveform into a binary sequence which also is an electric image of the pattern on the wheel. Again, different solutions can be used to convert the sinusoidal-like waveform into a binary sequence.

The most commonly used solutions are summarized in the following sections.

CHAPTER 4 – UNTREATED PROBLEM

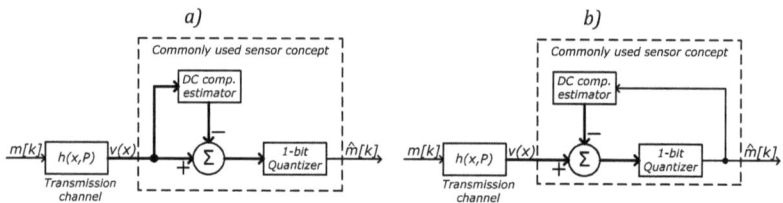

Figure 4.1: Commonly used sensor concepts
the DC component estimator and adder can work in the analog or digital domain
Estimating the DC component by evaluating the
a) analog waveform
b) binary sequence

4.1.1 Estimation of the DC Component

Estimating the DC component of a waveform (Figure 4.1.a) is a well known problem and a variety of different solutions exist. One solution is the use of a low pass filter that estimates the average value of the waveform [Bicking'94]. Other solutions calculate the average value of the waveform from its maxima and minima values [Infineon'05b]. The mean value of the last maxima and minima is approximately equal to the average value of the waveform. Other solutions for estimating the DC component of a waveform are presented in [Blossfeld'05, Moody'96, Moody'97b, Motz'04, Towne'01].

The value of the DC component can also be estimated by evaluating the binary sequence $\hat{m}[k]$ (Figure 4.1.b). If the binary sequence is continuously high or low, then a wrong DC value is subtracted from the analog waveform. The subtracted DC value can be varied until a switching of the binary signal is detected [Vig'04]. The DC value can be further adjusted until a 50% duty cycle of the binary output signal is reached [Blossfeld'05].

In the methods above, only one sensor element is used for measurements. A different approach involves two sensor elements which measure the magnetic field on two nearby angular positions of the wheel. The DC components of the analog output voltages of both sensor elements are similar to each other. The AC component is shifted by the separation distance of both sensor elements. If the separation distance is equal to the length of one region (magnetic region or tooth) on the wheel, subtracting these two signals minimizes the DC component and doubles the AC amplitude [Hainz'07a, Infineon'03]. Due to the imbalances between the two sensor elements, parasitic offset voltages etc., the DC component cannot be removed completely and 5%

to 10% remains [Vig'04]. Therefore, the above described strategies are additionally used to remove the remaining DC component.

4.1.2 Conversion of the Sinusoidal-Like Waveform in a Binary Sequence

A simple solution to convert the sinusoidal-like waveform in a binary signal is the use of a comparator which converts a value lower and higher than zero to a logic low and high state, respectively. The drawback of this solution is switching due to noise. Therefore, a Schmitt trigger circuit with hysteresis—which prevents switching due to noise—is more commonly used. However, the disadvantage of hysteresis is that the edges of the binary output signal are not exactly at the same time as the zero crossings. This delay of the binary signal results in angle measurement errors.

The disadvantage of hysteresis can be prevented by combining a comparator with a Schmitt trigger as presented in [Draxelmayr'02]. Once a zero crossing has occurred, the comparator switches. Further switching is disabled until the input signal reaches the threshold of the Schmitt trigger circuit. This combination prevents switching due to noise and allows switching on the exact positions of the zero crossings.

Detecting the zero crossings is not the only possible solution to convert the sinusoidal-like waveform into a binary sequence. The use of peak detection algorithms as shown in [Vig'94, Walter'98] delivers comparable results. However, the gradient of the AC waveform at the position of the peaks is small compared to the gradient at the position of the zeros. If peak detection is used instead of zero crossing detection, noise in the AC waveform causes more jitter in the binary sequence. On the other hand, peak detection is an AC function and therefore not influenced if the DC component is not removed completely [Walter'98].

Besides detecting zero crossings and peaks of the analog waveform, it is also possible to detect any other fixed levels between its minimum and maximum amplitude [Moody'96, Moody'97a, Towne'01]. However, in all of these cases, the sinusoidal-like waveform is converted in a binary sequence by using fixed level detection. As will be shown in the next section, the problem of commonly used sensor concepts cannot be avoided by changing the detection level.

4.2 Untreated Problem of Sensor Concepts in Use Today

Packaging and mounting tolerances, mechanical vibrations and temperature variations cause variations of the air gap between patterned wheel (pole wheel or toothed wheel) and sensor element. Due to this air gap variation, the magnetic field at the sensor position also varies resulting in variations of the electrical output signal of the sensor.

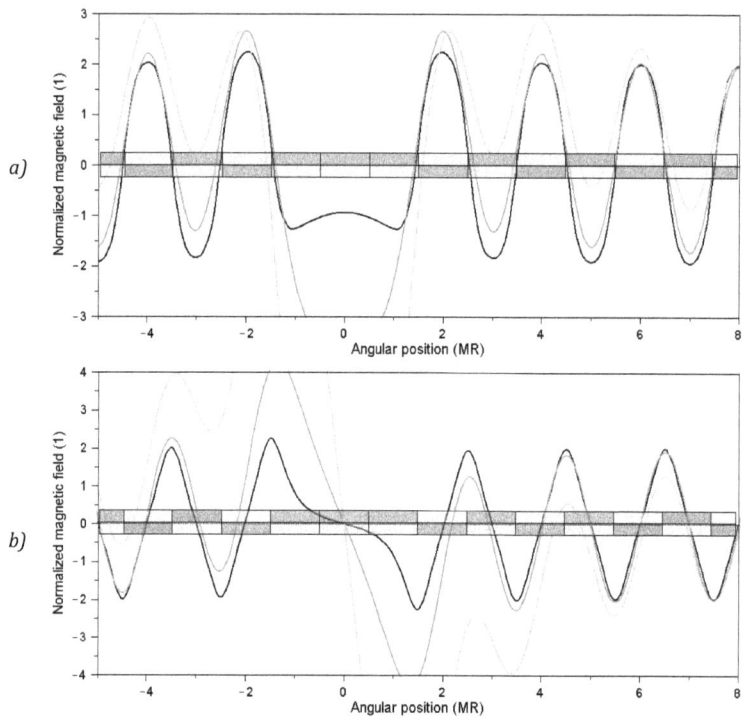

Figure 4.2: Magnetic field of a pole wheel (FEM simulations) with variations of the gap: 1.0 mm (dark), 2.5 mm; 4.0 mm (light) a) Normal component; b) Tangential component
[Scherhäufl'06]

Commonly used sensor concepts remove the gap-dependent DC component and use fixed level detection techniques to convert the remaining sinusoidal-like AC component in a binary sequence. However, they do not take into account the gap-dependent variations of the sinusoidal-like waveform shown in Figure 4.2

[Hainz'06c]. Gap variations cause a displacement of peak and zero crossing positions of the normal and tangential components of the magnetic field. Increasing the gap from 1.0 mm to 3.5 mm shifts the zero crossing position of the normal component by up to 0.28 MR and on a gap of 4.0 mm one zero crossing is even missing. Similar results are obtained detecting other fixed levels or evaluating the tangential component of the magnetic field.

Also on the sensing arrangement with a toothed wheel the sinusoidal-like waveform varies with the gap. Normalizing the AC component allows identifying the displacement of the zero crossing positions caused by gap variations (Figure 4.3.a). On the rotating toothed wheels eddy currents appear which cause a speed dependent displacement (Figure 4.3.b).

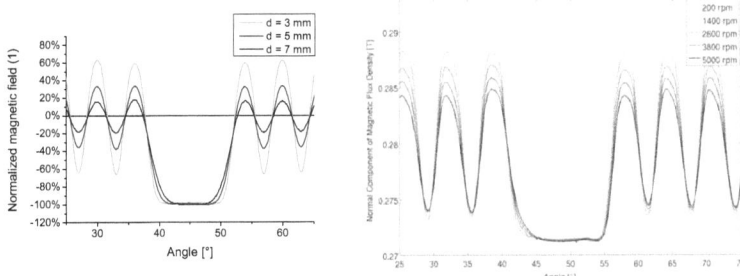

Figure 4.3: Magnetic field of a back-bias magnet, influenced by a toothed wheel (FEM)
a) Normalized field with variations of the gap: 3.0 mm (dark), 5.0 mm; 7.0 mm (light)
b) magnetic field at a gap of 2 mm with variations of the rotational speed
[Hainz'07a]

4.3 Reason for Research

It is expected that in the future, the demand for highest accurate crankshaft angle measurements will increase due to the requirements on engine emission reduction [Granig'07b]. Accurate measurements are also required for the so-called camless engine [Valeo'08].

The displacement caused by air gap variations is significantly larger than the accuracy of commonly used sensors. For example, a sensor in use today based on the Hall Effect is able to detect zero crossings with high accuracy and the 1 σ-jitter of the binary sequence is below 0.001 MR [Infineon'05a]. This sensor does not take into account that due to the gap variations, displacements of up to 0.28 MR of the zero crossing positions appears. Improving the sensor accuracy allows reducing the jitter by 0.001 MR whereby taking the displacement into account allows improving the accuracy by 0.28 MR. This makes clear that DC cancellation is an insufficient strategy in obtaining high angle accuracy.

Gap variations cannot be avoided; moreover, mechanical vibrations are expected to increase in the future due to the reduction of engine weight. Therefore, compensation of the gap-dependent displacement is the only solution to improve the angle accuracy significantly.

The high integration density of today's semiconductor technologies allows complex digital signal processing to be integrated on a small silicon area. Due to new technologies, high processing power is also available on the small chip area of integrated sensors. This processing power can be used for displacement compensation.

CHAPTER 5

Solution of the Problem

5.1 Introduction

The typical sensing arrangements for angle measurements based on magnetic sensing technique were shown in section 3.3. A patterned pole wheel or a toothed wheel is mounted on the rotating shaft and one or two sensors measure the magnetic field. This chapter focuses on the arrangement comprising a pole wheel and two sensor elements. Due to the similarities between the different arrangements, the solution discussed in this chapter can also be adapted for the use in the arrangement comprising a toothed wheel and a different number of sensor elements.

5.1.1 Reason for the Displacement of the Zero Crossing Positions

If we look at this sensing arrangement from a digital communication theory point of view, we have a transmission channel $g_M(x,\mathbf{P})$ that transforms the pattern $m[c]$ along the circumference of the pole wheel to a magnetic field at the position of the sensor element (Figure 5.1). The magnetic field at the position of the sensor elements depends primarily on the angle of the wheel and secondarily on physical parameters $\mathbf{P}$ like the gap between sensor element and patterned wheel [Hainz'06d].

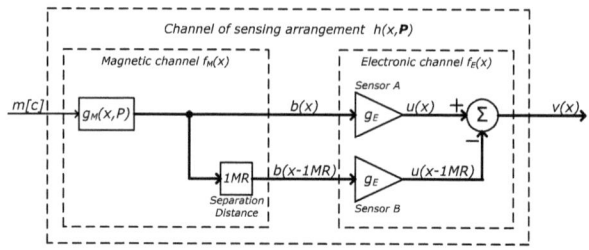

Figure 5.1: Channel of the differential sensing arrangement with
m[c]: magnetic pattern along the circumference of the pole wheel;
$b(x)$ and $b(x-1\text{MR})$: magnetic field at the position of sensor element A and B;
$u(x)$ and $u(x-1\text{MR})$: electrical variables of the sensor elements;
$v(x)$: differential output voltage;

The sensor element varies an electric variable in response to the changes in the applied magnetic field. (Hall elements vary their output voltage and GMR elements their resistance, see section 3.3.3.) The two sensor elements are typically placed on a separation distance which is equal to the length of one magnetic region and measure the magnetic field at the positions x and $x-1\text{MR}$, respectively [Infineon'03].

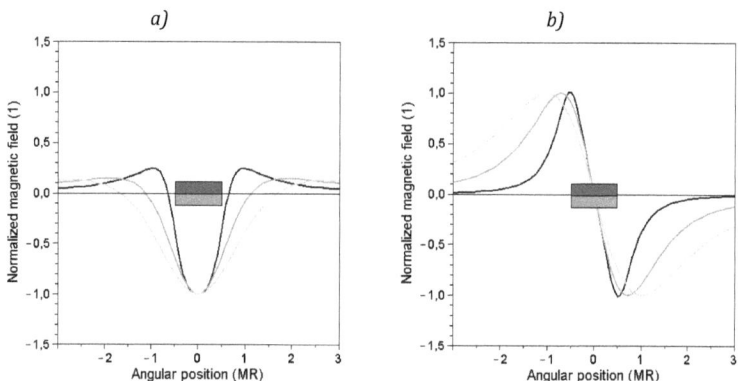

Figure 5.2: Pulse shaped magnetic field generated by a single magnetic region
with variations of the gap: 1.0 mm (dark), 2.5 mm; 4.0 mm (light)
a) Normal component; b) Tangential component

The differential output voltage $v(x)$ can be calculated from the electrical variable of the two sensor elements by using analog circuits such as a differential amplifier or the Wheatstone bridge. Subtracting the output $u(x-1\text{MR})$ of sensor element B from the output $u(x)$ of element A leads to the differential output voltage

$$v(x) = u(x) - u(x-1\text{MR}). \qquad (5.1)$$

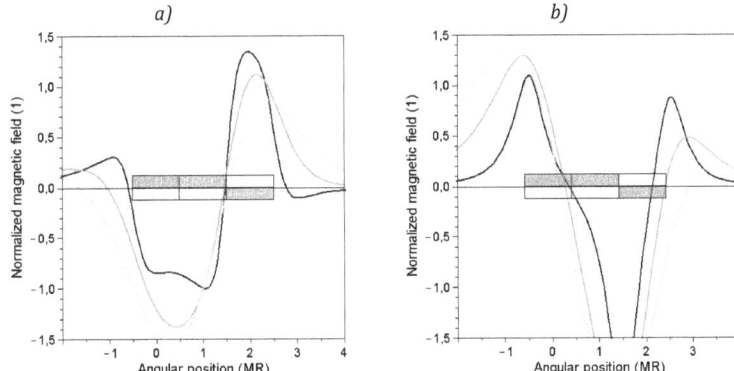

Figure 5.3: Magnetic field distribution generated by three magnetic regions with variations of the gap: 1.0 mm (dark), 2.5 mm, 4.0 mm (light) a) Normal component; b) Tangential component (calculated using linear superposition)

Transmitting a single magnetic region through the magnetic channel $h(x,\mathbf{P})$ causes a magnetic field at the position of the sensor element with a pulse shaped normal component as shown in Figure 5.2.a. The transfer characteristic of the magnetic channel depends on the physical parameters $\mathbf{P}$. Varying the physical parameter values, results in variations of the pulse shape. Increasing the gap causes a lowering of the magnetic field intensity (pulse peak) and an increasing pulse-width. The lowering of the field cannot be seen in Figure 5.2 due to normalization.

In contrast to the pulse shape of the normal component, the shape of the tangential component is a double-pulse as shown in Figure 5.2.b. Again, increasing the gap causes a decreasing magnetic field and an increasing width.

Transmitting more than one magnetic region and assuming that each one generates such a field, the resulting field $b(x)$ can be calculated using linear superposition [Hainz'07a, Hainz'06d, Kajley'97]. Calculation results—adding three pulses—are shown in Figure 5.3. The magnetic field generated by each region extends to the neighboring regions. The field, generated by neighboring regions, interferes with each other. Because the pulse-width varies as a function of the gap, the interference also varies as a function of the gap resulting in a gap-dependent displacement of the peak and zero crossing positions.

Depending on the pattern of the three regions, the interference of the two outer regions to the inner one can cause a displacement in the same or opposite direction.

The value and direction of the cumulative displacement does not only depend on the gap, but also on the pattern on the wheel.

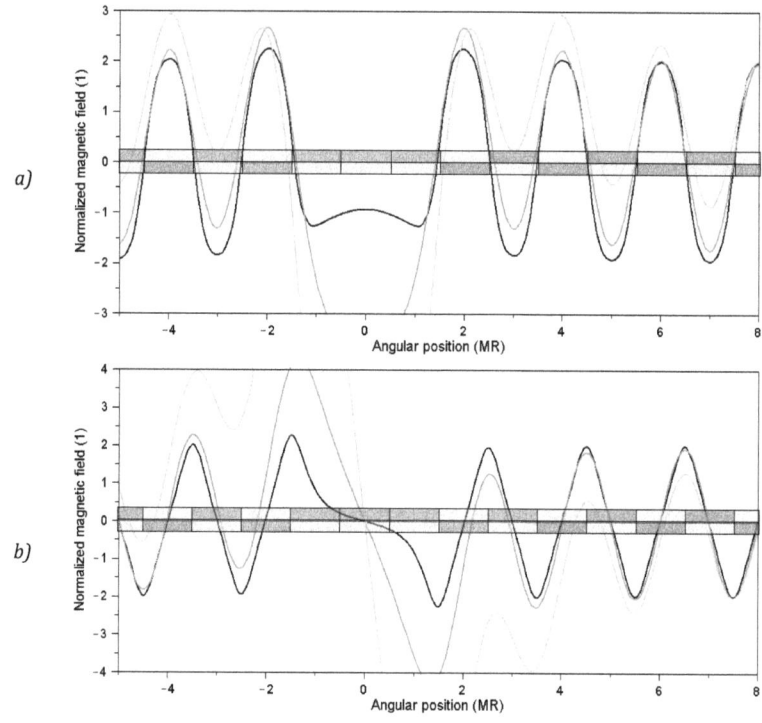

Figure 5.4: Magnetic field of a pole wheel (linear superposition) with variations of the gap: 1.0 mm (dark), 2.5 mm, 4.0 mm (light)
a) Normal component; b) Tangential component

The magnetic field of a pole wheel with 120 alternating magnetic regions along its circumference and one synchronization point was calculated using linear superposition (Figure 5.4). The linear superposition results match the finite element method (FEM) simulation results presented in section 3.3.1 and the deviations between both are below 2% on gaps between 1 mm and 6 mm.

The gap-dependent displacement of the zero crossing positions can be observed using both FEM simulations and linear superposition. Using linear superposition, the interference between neighboring regions was identified as the reason for the displacement.

5.1.2 Intersymbol Interference in Other Applications

Interference of neighboring regions is also known in other applications such as the replay process of a hard disk drive. A magnetic field sensor measures the magnetic field above a magnetized rotating platter [Bergmans'96, Kajley'97]. Each magnetic region on the platter generates a pulse shaped magnetic field. High data densities move the magnetic regions closer to each other and interference between the fields of neighboring regions appears. In literature this interference is called intersymbol interference (ISI).

ISI is also known in the area of digital baseband transmission (e.g., wired communication) where voltage pulses are transmitted over a transmission channel (e.g., wire). Typically the baseband transmission channel has a frequency response of a low pass filter. Different frequency components of the transmitted signal are attenuated and removed by the channel [Barry'04] resulting in a spreading of the pulses. The pulses at the receiver are broader than the transmitted pulses and they interfere with each other.

In wireless communication the signal from the transmitter reaches the receiver via many different paths which are caused by the reflection (e.g., buildings, ionosphere) and refraction (e.g., tropospheric turbulences) of the electromagnetic wave on different objects (e.g., mountains, houses). The different paths vary in their length and damping and therefore many different time-shifted and attenuated versions of the signal arrive at the receiver. Due to the delay between the signals each symbol spreads to the next symbols. In contrast to wired communication, where the channel characteristic is constant over time, in wireless communication the channel characteristic varies [Treichler'01].

The above examples have shown that ISI is well known in different areas. Also if the transfer characteristic of the transmission channel differs between applications, the effect of ISI still remains the same. Therefore known techniques to deal with ISI in different applications were analyzed for the use in sensor applications.

In wired communications for example, where the transfer characteristic $g_c(t)$ is equal to a low pass filter, ISI can be avoided or removed by a mapper, a filter at the sender or a filter at the receiver (Figure 5.5) [Barry'04].

The mapper maps one or more data bits $a[i]$ into a symbol $b[j]$. The frequency spectrum of the symbol sequence can be higher or lower compared to the spectrum of

Chapter 5 – Solution of the Problem

the data sequence which allows matching the spectrum of the symbol sequence to the spectrum of the channel. ISI is avoided if the spectrum of the symbol sequence is matched to the spectrum of the channel [Barry'04].

ISI caused by the channel can also be avoided by limiting the bandwidth of the transmitted symbols to the bandwidth of the transmission channel. This can be done using the filter at the sender.

The third possibility to deal with ISI is the use of a filter at the receiver. For example, the transfer characteristic of this filter can be chosen to force ISI to zero. Other filter strategies are discussed in the following section.

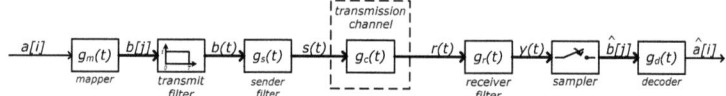

Figure 5.5: General baseband transmission channel using pulse amplitude modulation (PAM)

5.1.3 Known Techniques to Deal with ISI

On constant revolution speed, the channel of the sensing arrangement in the space domain (Figure 5.1) can be transposed into the time domain (Figure 5.6) and the presented techniques to deal with ISI can be applied to the channel of the sensing arrangement. However, the channel of the sensing arrangement cannot be modified and therefore the use of a mapper and a filter at the sender is not possible. The only possible solution to deal with ISI in sensor applications is the use of a filter at the receiver. The three commonly used filter types in digital communication are summarized in this section.

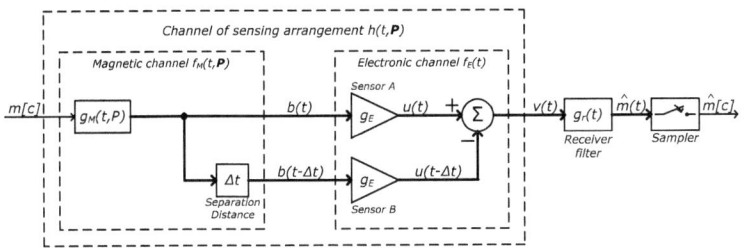

Figure 5.6: *Transmission channel of the sensing arrangement in the time domain*
m[k]: *magnetic pattern along the circumference of the pole wheel;*
$b(t)$ and $b(t-\Delta t)$: *magnetic field at the position of the sensor element A and B;*
$u(t)$ and $u(t-\Delta t)$: *sensor elements output voltage;*
$v(t)$ *differential output voltage;*

Partial response maximum likelihood (PRML) is a technique to convert an analog waveform into the corresponding sequence of symbols by interpreting even small changes in the signal. The PRML detector compares the measured analog signal at the receiver with the calculated waveforms for all possible symbol sequences. Then the sequence whose waveform has the best match to the measured signal is chosen as decoded symbol sequence [Barry'04]. The complexity of a PRML receiver increases exponentially with the length of the considered sequence of symbols. If the channel has a long impulse response, a long sequence must be considered. Therefore, simpler techniques such as the linear equalizer and the decision feedback equalizer are more commonly used [Barry'04].

The linear equalizer employs the filter $g_r(t)$ at the receiver and removes ISI of the channel. Two design procedures, minimizing the ISI or the mean square error, are commonly used and summarized in chapter 2. The optimal transfer characteristic of $g_r(t)$ to remove ISI completely, is the reciprocal value of the channel transfer characteristic. Using this so-called zero forcing equalizer, ISI can be completely removed; however, removing ISI enhances noise [Barry'04, Le'02]. If ISI is removed completely, noise enhancement can be infinite [Bergmans'96]. A tradeoff must be found between noise enhancement and ISI cancellation [Barry'04]. A different approach, minimizing the mean square error (i.e., the compound of ISI and noise) at the receiver, avoids infinite noise enhancement but ISI is no longer removed completely.

The decision feedback equalizer (DFE) is based on the linear equalizer and combines low complexity with high performance. A DFE consists of a feed forward filter and a feedback filter. The feed forward filter, a conventional linear equalizer, cancels the interference of future symbols. Instead, the feedback filter evaluates the decoded symbols and calculates the interference of these symbols. Then the interference of the decoded symbols is subtracted from the output of the feed forward filter [Barry'04, Bergmans'96]. Compared to a PRML detector, the complexity of a DFE is small and its performance can be equal or even better [Brown'99, Kajley'97, Rothenberg'97]. The noise enhancement of the DFE is small compared to the linear equalizer because the ISI of future symbols is cancelled by the feedback filter without noise enhancement [Bergmans'96, Spencer'99].

5.1.4 Aim of the Thesis

The aim of this thesis is to examine whether digital signal processing can be used to improve the angle accuracy of the sensing arrangement. It was shown that the systematic angle error is caused by ISI. Each of the three filter types summarized in section 5.3.1 deals with ISI and can be used to improve the angle accuracy. However, the DFE combines small complexity with high performance and is therefore preferred to the other filter types.

The digital filter, developed in this thesis, is based on the DFE. The realization consists of three parts (Figure 5.7):

- *Displacement Compensation*
  The DFE was modified for use in sensor application. In the new filter structure the feed forward filter is replaced by an additional filter in the feedback path. The complexity of the modified DFE is small compared to the complexity of the conventional DFE.
- *Filter Coefficients Estimation*
  An adaptive algorithm estimates the filter coefficients of the modified DFE. Rather than estimating the filter coefficients directly, a simple physical model of the sensing arrangement is used and the values of the physical parameters are estimated.
- *Clock Signal Generation*
  To reduce the complexity of the DFE and the adaptive algorithm, the clock frequency must be synchronous to the revolution speed. An enhanced PLL, which predicts the phase error, can follow the crankshaft speed fluctuations and generates the clock signal.

In the following sections each part of the digital filter is presented separately and simulation results explain its functionality. Simulation results of the complete filter structure and performance limit calculations are shown in section 5.5.

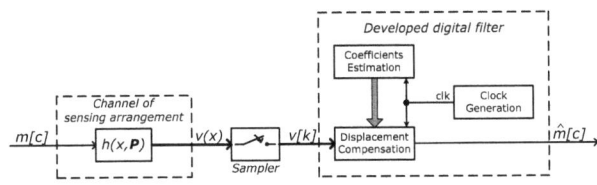

Figure 5.7: Block diagram of filter structure

5.2 Displacement Compensation

The modified DFE, presented in this section, can be used to remove the intersymbol interference caused by the transmission channel of the magnetic sensing arrangement. In this chapter it is assumed that the optimal filter coefficients are known. It is also assumed that the data rate of the sender (i.e., the rotational speed of the wheel) is known at the receiver side.

The similarities between the magnetic transmission channel (Figure 5.6) and the partial response filter with transmission channel (Figure 2.6) make clear that the DFE can be used to remove ISI resulting from the magnetic transmission channel if the sampling period of DFE is equal to the delay Δt between the signals of the two sensor elements.

5.2.1 Increasing Angle Accuracy

Using the DFE to remove the ISI of the transmission channel of the sensing arrangement allows determining the magnetic states of the pole wheel. In contrast to digital communication, in sensor applications the required information is the transition between magnetic states (position) and not the states (data) itself. Therefore the pattern on the wheel is assumed with higher symbol density (Figure 5.8). Determining the magnetic states of the high density pattern enables identifying the transition between magnetic states more accurately. [Hainz'04a]

CHAPTER 5 – SOLUTION OF THE PROBLEM

At higher symbol density, the DFE has to work at higher sampling rate $f_k = \psi \cdot f_c$. To remove the same amount of ISI (e.g., ISI over one magnetic region) at higher sampling rate, the filter order has to be increased by factor ψ.

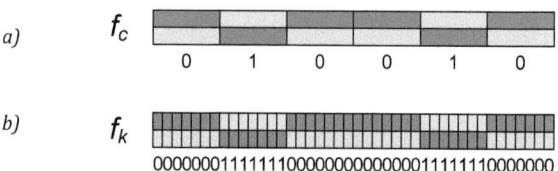

Figure 5.8: Magnetic pattern along the circumference of the pole wheel
a) magnetic pattern m[c] with symbol density f_c
b) magnetic pattern m[k] with assumed higher symbol density $f_k = \psi \cdot f_c$

At a high sampling rate the input signals for the feed-back filter (FBF) and feed forward filter (FFF) are the binary sequence $\hat{m}[k]$ (equals $\hat{m}[c]$ with higher sampling rate f_k) and the multi-bit signal $v[k]$, respectively. Increasing the filter order for both filters causes a moderate increase in filter complexity for the single-bit FBF and a drastic increase for the multi-bit FFF. To reduce complexity of the DFE, a structure without need of a FFF is required.

5.2.2 Predictive Feedback Filter

A modified filter structure was introduced in [Hainz'04a] and called »Predictive DFE« (pDFE). This structure works at high sampling rate f_k and employs a conventional FBF (now called post-FBF) to generate and remove post-ISI. The FFF is replaced by an additional filter in the feedback path, called pre-FBF (Figure 5.9).

In angular measurements the rotating patterned wheel generates a periodic sequence m[k]. If the pattern and the angular position of the wheel are known, future »data« (i.e. the magnetic states of the next regions) can be predicted. The additional filter in the feedback path of the pDFE (called pre-FBF) uses the predicted magnetic states of the next regions to calculate their interference to the current position (pre-ISI). Then the pre-ISI is subtracted prior to discrimination.

Chapter 5 – Solution of the Problem

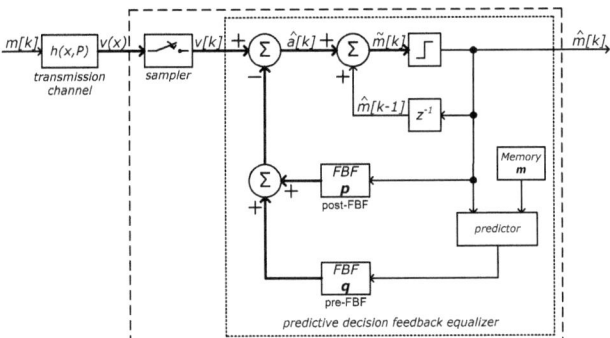

Figure 5.9: Predictive decision feedback equalizer

Typical waveforms of a pDFE can be seen in Figure 5.10.a for a gap of 1.0 mm. The generated pre- and post-ISI is subtracted from the input signal and only the differential data $â[k]$ (i.e. the symbols) remain. The partial response equalizer (section 2.1.3) converts the differential data into the sequence $\hat{m}[k]$.

The same filter structure can be used to remove ISI on different air gaps (e.g., 3 mm Figure 5.10.b) if filter coefficients are varied. The remaining symbols $â[k]$ after ISI cancellation are independent on the air gap. Small differences in the quantization noise can be observed, which is caused by the varying slope of the signal $v[k]$ (this effect is discussed in more detail in section 5.5. However, the noise in $â[k]$ is removed by the slicer.

CHAPTER 5 – SOLUTION OF THE PROBLEM

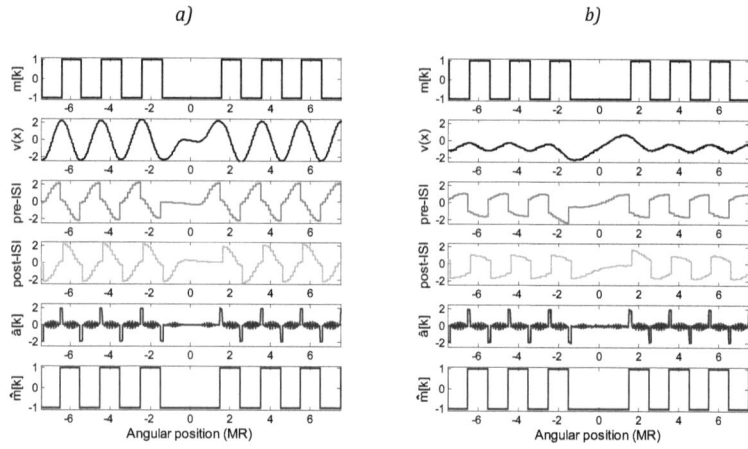

Figure 5.10: Simulation result of the pDFE for a GMR sensor element with variations of the gap between sensor element and pole wheel
a) 1.0 mm
b) 3.0 mm

5.2.3 Start-Up Phase

The presented pDFE removes ISI if the pattern on the pole wheel and the angular position is known. Therefore a start-up phase—learning the pattern and determining the position—is required. [Hainz'04b]

Peak or zero crossing detection of signal $v(x)$ can be used to generate the binary sequence $\hat{m}(x,t)$ which is a rough estimation of the pattern on the wheel (Figure 5.11). Assuming that each magnetic region on the pole wheel has the same length (e.g., 2.5 mm), the pattern on the wheel also can be determined if ISI appears. Knowing the pattern, cross-correlation can be used to determine the position.

Once the pattern is learned and the position determined, the demultiplexer of the pDFE switches into running mode. In the running mode the ISI of the channel is removed and the output signal $\hat{m}[k]$ is independent on the air gap.

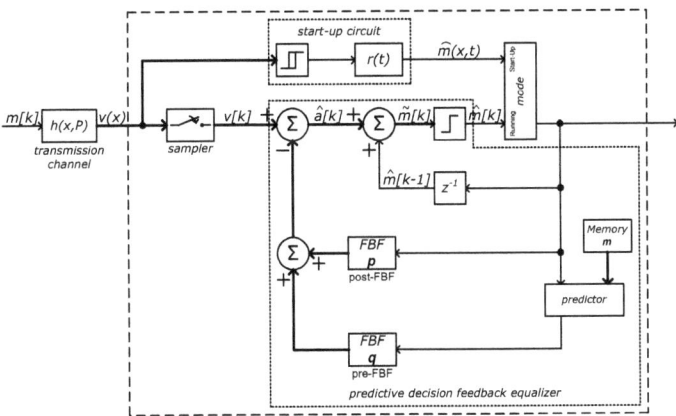

Figure 5.11: pDFE with start-up circuit

5.2.4 Implementation Hints

5.2.4.1 Complexity reduction of the Feedback Filter

Sensor concepts in use today detect zero crossings with an accuracy in the range of 0.01 MR [Infineon'03]. To achieve the same accuracy using the DFE, the sampling rate and filter order must be increased by a factor of 100. To remove ISI over one magnetic region a FIR filter with an order of 100 is required for each, the pre- and post-FBF. Complexity of the pDFE can be reduced by replacing the FIR filters with look up tables (LUTs) and counters.

The data sequence $\hat{m}[k]$ is equal to the oversampled magnetization pattern $\hat{m}[c]$ by a factor of 100. Therefore the sequence $\hat{m}[k]$ consists of sub-sequences of 100 successive logic high or low states. Using this sequence as input for the FIR filter, the sequence of 100 logic high or low states is shifted through the filter. The 100 unit delays of the FIR filter store the information of the transition point between two magnetic states. This time information can be measured easier by a counter (also called time to digital converter) which starts counting with a rising or falling edge of the input signal $\hat{m}[k]$ (Figure 5.12). Then the counter value is used as index for a LUT. The stored information in the LUT (e.g., the values of the post-ISI) is read out successively.

Using the new structure, reduces the complexity of the pre- and post-FBF. Additional benefits are the reduction of processing power (multipliers and adders of a FIR filter) and consequently the decrease in power consumption.

CHAPTER 5 – SOLUTION OF THE PROBLEM

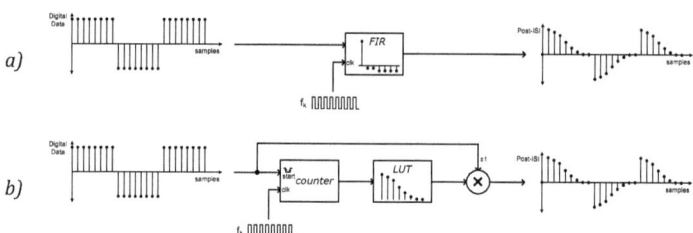

Figure 5.12: Replacement of the post-FBF by a counter and LUT with $\psi = 8$
a) conventional FIR filter
b) counter and look up table

Removing ISI over more than 1 MR, the value of the post-ISI at the current position depends on the time information of more transitions between magnetic states. Therefore the use of multiple counters and LUTs in parallel, as shown in Figure 5.13, is proposed.

Multiple counters and LUTs generate the post-ISI of multiple magnetic regions. The post-ISI of the magnetization pattern $m[k]$ is generated by adding the output signals of the LUTs.

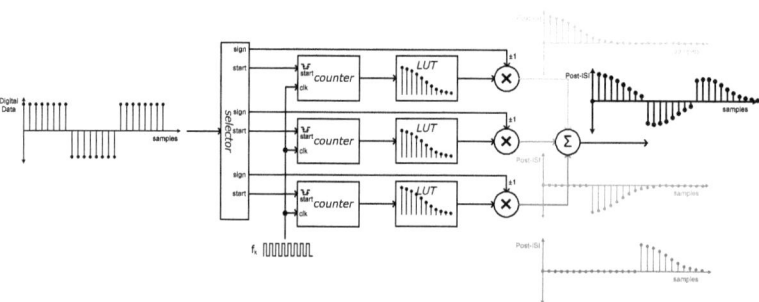

Figure 5.13: Replacement of the post-FBF by multiple counters and LUTs with $\psi = 8$

5.2.4.2 Complexity reduction of the Start-Up Circuit

As described in section 5.2.3, cross-correlation can be used to determine the position of the learned magnetization pattern $m[c]$ within the periodic sequence $\hat{m}(x,t)$. Assuming a pole wheel with 120 magnetic regions, cross correlation between the 120 by 1 matrix $\mathbf{m}_{1-120,1}$ of the learned pattern $m[c]$ and the 1 by 120 matrix $\hat{\mathbf{m}}_{1-120,1}$ of the last 120 measured magnetic states $\hat{m}(x,t)$ on the wheel must be calculated which requires high processing power.

-50-

Processing power can be reduced if a counter measures the pulse length of the logic high and low states in the sequence $\hat{m}(x,t)$. Using a counter, the pattern on the pole wheel is expressed as a sequence of integer values. This sequence of integer values can be searched for a characteristic of the pattern (e.g., the longer synchronization point) and the position can be determined.

5.2.5 Complexity Estimation

To estimate the complexity of the pDFE, the resolution is defined as 0.01 MR (i.e., $\psi = 100$) and ISI should be removed over 4 MR. Two FIR filters with 400 coefficients are required to remove ISI over 4 MR with a resolution of 0.01 MR.

With a factor $\psi = 100$, the FIR filters have to work on a frequency of $f_k = 100 \times f_c$. If the maximal rotational speed is defined as 9,000 rpm, the FIR filters work with a sampling rate below $f_{k_{max}} = 900\,\text{kHz}$.

Complexity can be reduced by replacing each FIR filter by a counter with maximal counter value of 400 and a LUT. If the bit width of the multi-bit signals is equal to 12 bit, a 400×12 bit LUT is required.

The start-up circuit requires a counter to express the pattern on the wheel as a sequence of integer values, a 120 bit memory to store the sequence and some computational power to search the characteristic of the pattern within the stored values in the memory.

The implementation of the digital filter with start-up circuit requires three 9-bit counters and one 410×12 bit memory. The maximal sampling rate of the counters and memory is equal to 900 kHz.

5.3 Adaptive Algorithm using Physical Model of Sensing Arrangement

If the transfer characteristic of the channel is not known, adaptive algorithms are used to estimate the filter coefficients of the DFE. Due to the similarities between the predictive decision feedback equalizer (pDFE) and the DFE, the same adaptive algorithms can be used on both structures (Figure 5.14).

CHAPTER 5 – SOLUTION OF THE PROBLEM

In digital communications ISI typically is removed over a few symbols [Brown'96, Kajley'97] resulting in a DFE with low order (e.g., 2 to 3) FIR filters. The performance function of a second order FIR filter is three-dimensional (error signal versus its filter coefficients b_0 and b_1) and the search for local minima can be performed easily.

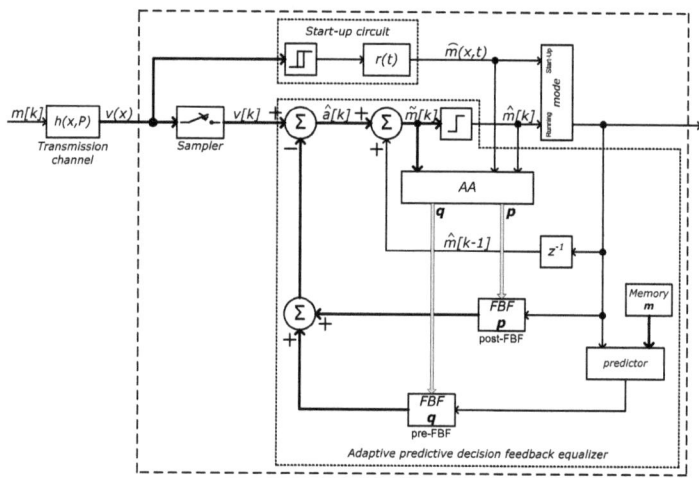

Figure 5.14: pDFE with adaptive algorithm

In sensor applications high angle accuracy requires high data rates resulting in high order (e.g., 400) FIR filters. Using hundreds of uncorrelated filter coefficients causes a hundreds-dimensional performance function and the search for local minima will develop to a highly complex problem. To reduce complexity an adaptive algorithm with reduced degrees of freedom is required.

The second problem in sensor applications is that the training sequence (the phase of the sequence $m[k]$) is not available. During the start-up phase (section 5.2.3) zero crossing detection is used to generate the binary sequence $\hat{m}(x,t)$ which is a rough estimation of the pattern on the wheel. Using $\hat{m}(x,t)$ as a low quality training sequence allows the AA to coarsely estimate the filter coefficients and ISI is partly removed. After adaption is completed the angle accuracy of the pDFE is equal to the accuracy of the binary sequence $\hat{m}(x,t)$. Further improvement of the angle accuracy is not possible because many local and global minima appear in the hundreds-dimensional performance function. Simulations have shown that the filter coefficients can be wrongly estimated even if the error signal is minimized.

CHAPTER 5 – SOLUTION OF THE PROBLEM

Both problems can be solved by reducing the adaptive algorithm degree of freedom. Using a physical model of the measurement setup, the algorithm estimates a few physical parameters instead of hundreds of filter coefficients.

5.3.1 Estimation of the Physical Parameter Values

The transmission channel between pole wheel and sensor element can be described by the transfer function $h(x,\mathbf{P})$ which depends on the physical parameters $\mathbf{P}$ and the angle of the patterned wheel. If the relationship between the physical parameters and transfer function $h(x,\mathbf{P})$ is known, the transfer characteristic (distortion) of the channel can be calculated by the values of the physical parameters [Hainz'06d].

Using this approach the adaptive algorithm in Figure 5.15.a can be extended as shown in Figure 5.15.b and the values of the physical parameter are estimated instead of estimating the filter coefficients directly. With the adapted physical values the channel distortion—and consequently the filter coefficients to remove this distortion—can be calculated using the physical model. [Hainz'08a]

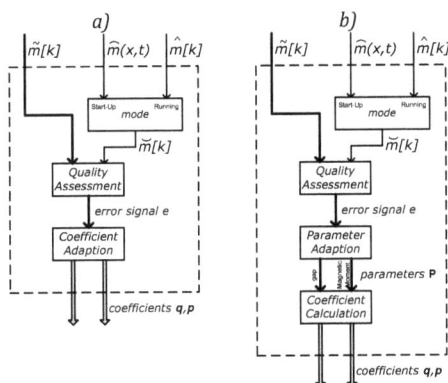

Figure 5.15: Adaptive algorithm of the DFE and pDFE
a) estimation of all filter coefficients separately
b) estimation of the physical parameters

The few-dimensional performance function has no local and only one global minimum (see section 5.3.4). Therefore the binary sequence $\tilde{m}(x,t)$ can be used as a (low quality) training sequence which allows the adaptive algorithm to coarsely estimate the physical parameters and roughly find the global minimum in the few-dimensional performance function. Then the pDFE switches in running phase and uses the signal

Chapter 5 – Solution of the Problem

$\hat{m}[k]$ as the new training sequence. Starting from the previous coarsely estimated parameters, their value now can be estimated more precisely. The phase accuracy of the system is improved successively.

5.3.2 Simple Physical Model of Sensing Arrangement using Pole Wheel

Estimating a few physical parameter values instead of hundreds of uncorrelated filter coefficients requires less processing or hardware for the adaptive method [Hainz'08a]. However, further processing or hardware is required to calculate the filter coefficients from the physical values. Due to the limited calculation power of the sensor chip, a simple physical model for the transmission channel must be found.

The magnetic field generated by one atom of a permanent magnet is equal to the field generated by an electrical current flowing through an infinitesimal loop [Fitzpatrick'07, Fricke'86, Oberdorfer'61, Preis'04]. Modeling each atom of the magnet by an infinitesimal loop with the same magnetic moment leads to the model shown in Figure 5.16.a [Fitzpatrick'07]. The interior currents flow in opposite directions and cancel each other out. Only the current on the surface of the magnet remains (Figure 5.16.b). Also in the three-dimensional space, the magnetic field of a permanent magnet is equal to the field generated by a solenoid with the same dimensions and magnetic moment (Figure 5.17). However, the field of a permanent magnet and a solenoid is only equal at the outside of the magnet (i.e. in vacuum) [Fitzpatrick'07, Fricke'86, Oberdorfer'61, Preis'04].

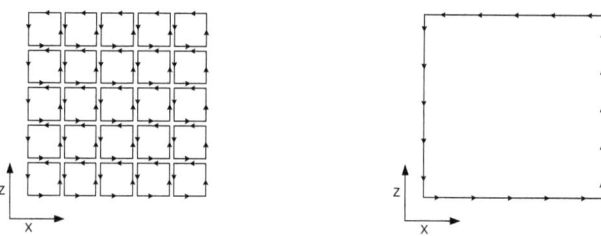

Figure 5.16: Cross section of a permanent magnet
a) modeling each atom by an infinitesimal loop with current
b) remaining current at the outer surface of the magnet

To reduce the necessary computing power to calculate the magnetic field, the solenoid is assumed with infinitesimal length and infinite depth. The magnetic field of the permanent magnet is approximated by the field of a current flowing through two wires in opposite directions (Figure 5.17.c).

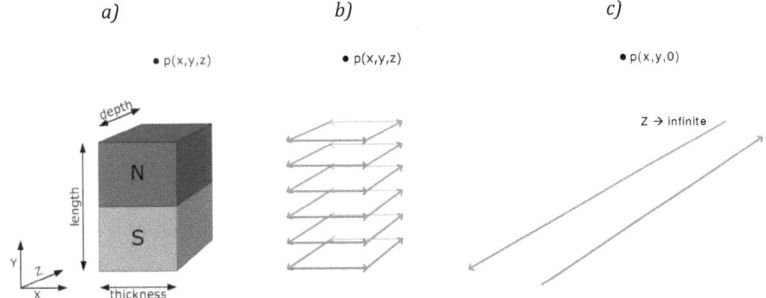

Figure 5.17: Simple model of a permanent magnet
a) permanent magnet
b) solenoid with same dimensions and magnetic moment as the permanent magnet
c) two wires with current

The normal and tangential component of the magnetic field at the position $p(x,y,0)$ generated by a wire with current I is given by

$$B_x = \frac{\mu_0 \cdot I \cdot y}{2 \cdot \pi \cdot (x^2 + y^2)} \tag{5.2}$$

$$B_y = \frac{-\mu_0 \cdot I \cdot x}{2 \cdot \pi \cdot (x^2 + y^2)}, \tag{5.3}$$

respectively, where μ_0 is the vacuum permeability and I is the current through the wire [Preis'04]. The magnetic field of a permanent magnet can be approximated by

$$B_x = \frac{\mu_0 \cdot I \cdot y}{2 \cdot \pi \cdot \left((x+0.5 \cdot x_{thick})^2 + y^2\right)} + \frac{-\mu_0 \cdot I \cdot y}{2 \cdot \pi \cdot \left((x-0.5 \cdot x_{thick})^2 + y^2\right)} \tag{5.4}$$

$$B_y = \frac{-\mu_0 \cdot I \cdot (x+0.5 \cdot x_{thick})}{2 \cdot \pi \cdot \left((x+0.5 \cdot x_{thick})^2 + y^2\right)} + \frac{\mu_0 \cdot I \cdot (x-0.5 \cdot x_{thick})}{2 \cdot \pi \cdot \left((x-0.5 \cdot x_{thick})^2 + y^2\right)}, \tag{5.5}$$

where x_{thick} is the thickness of the magnet. To reduce computing power, the rotation of the wheel was modeled as linear movement of the sensor element over an infinite magnetic strip with repeating magnetic pattern (Figure 5.18). Replacing each magnet with two wires with current I the magnetic field at the position $p(x,y,0)$ can be calculated using linear superposition [Fricke'86, Oberdorfer'61]. Some currents flow in opposite direction and cancel out each other (Figure 5.18).

CHAPTER 5 – SOLUTION OF THE PROBLEM

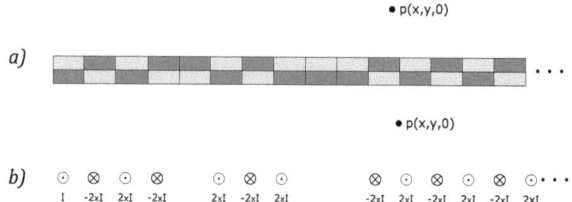

Figure 5.18: Wire model of a magnetic strip
a) magnetic strip; b) wire model

The above presented simple wire model of the measurement setup has two parameters:
1. The air gap »G« between sensor element and pole wheel,
   which is proportional to y in (5.4) and (5.5)
2. The magnetic dipole moment »M« of the permanent magnet,
   which is proportional to I in (5.4) and (5.5)

The output of the sensor element (Hall, GMR) is assumed to be linear in proportion to the applied magnetic field.

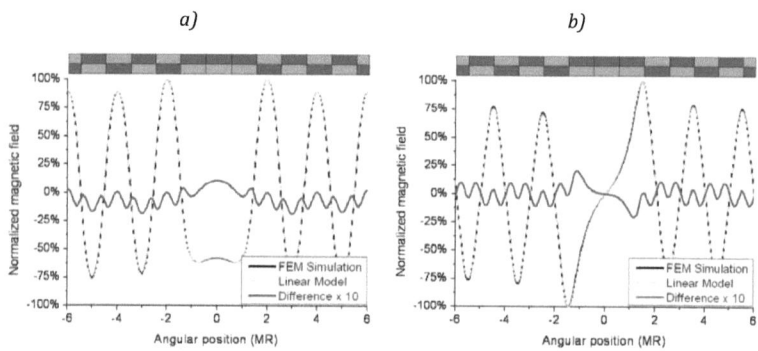

Figure 5.19: Magnetic field of a pole strip at an air gap of 1 mm
FEM simulation results (dark blue), calculations using wires model (dashed light blue)
a) normal component
b) tangential component

FEM simulations were performed to verify the wire model. The magnetic field of a magnetic strip was compared with the field of its corresponding wires model. The deviation between FEM simulation results of the magnetic strip and calculation results

-56-

of the wires model decrease with the air gap and are below 2% on gaps between 1 mm and 6 mm (Figure 5.19) [Hainz'07a, Scherhäufl'06, Spencer'99].

Further, the radius of the wheel was taken into account and the FEM simulation results of a pole wheel were compared with the linear wire model shown in Figure 5.18. The deviation between FEM simulation and calculation increases with the gap and decreases with the radius. On a pole wheel with a radius of 50 mm, the deviation is 3% and 10% at an air gap of 1 mm and 6 mm, respectively. The results of FEM simulation are shown in more detail in [Scherhäufl'06].

5.3.3 Error Signal

An error signal—which is a measure of the quality of the guessed parameter values—is required for the adaptive method.

5.3.3.1 Summation of the Squared Difference

The squared difference between the signal $\tilde{m}[k]$ (the signal after partly removing ISI) and the training sequence $\bar{m}[k]$ i.e.,

$$e[k] = (\tilde{m}[k] - \bar{m}[k])^2 \tag{5.6}$$

identifies remaining ISI in $\tilde{m}[k]$.

ISI can remain due to a wrong estimate for the parameter M, G or both. To calculate the next guess, the information which one of the parameters must be changed is essential. Therefore the summation of the squared difference

$$\varepsilon = \sum e[k] = \sum (\tilde{m}[k] - \bar{m}[k])^2 \tag{5.7}$$

is chosen as error signal. Now, the waveforms of both signals are compared instead of comparing the values at a single point in time. Comparing the waveforms, the information which one of the parameters must be changed can be estimated.

The summation length of (5.7) can be freely chosen, but with a too small summation length the parameter values can only coarsely estimated. Therefore a tradeoff must be found between accuracy and speed.

5.3.3.2 Characteristic Parameters

A second approach is to determine specific characteristic parameters (amplitude, pulse-width, etc.) of both waveforms and to calculate the next guess by comparing the values of the characteristic parameters. The pulse-width at the position of the longer magnetic region increases with the air gap. The air gap can be estimated by adjusting the pulse-width of the signal $\tilde{m}[k]$ to the pulse-width of $\bar{m}[k]$. However, on wheels with other patterns the pulse-width can increase or decrease with the air gap. Therefore this approach can only be used if the physical parameter values vary over a small range or on wheels with specific patterns. In this thesis the second approach is not discussed in more detail.

5.3.4 Performance Function and Learning Function

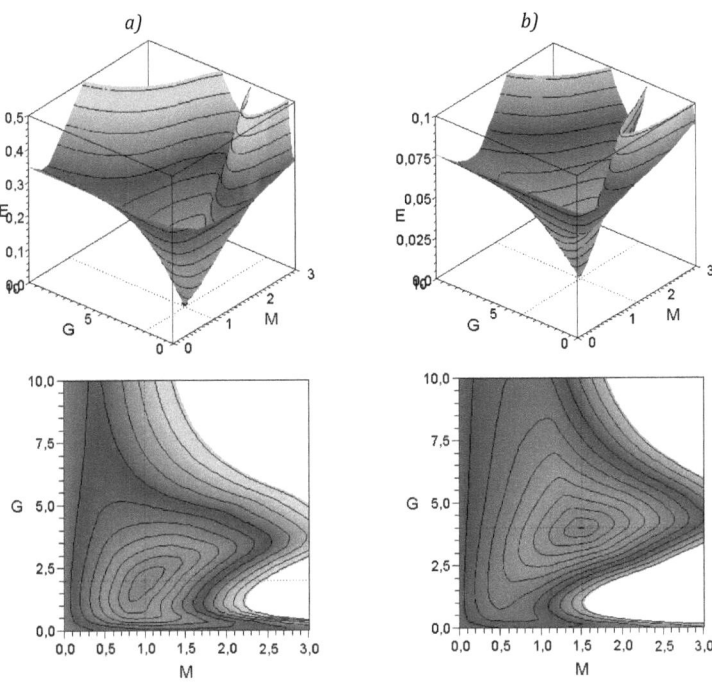

Figure 5.20: Performance function of the adaptive algorithm using physical model with variations of the channel parameter values
a) G = 2.0, M = 1.0
b) G = 4.0, M = 1.5
G = air gap, M = magnetic dipole moment

The modeled sensing arrangement consists of a pole wheel with one synchronization point (Figure 3.2.a) and GMR elements measure the magnetic field at two positions on the wheel. The summation length of the error signal ε (expressed as geometric length) was chosen as 25 MR.

Calculations were carried out in Maple 11 for different values for the physical parameters of the transmission channel. The resulting performance functions (Figure 5.20) have their global minima at the position where the estimated values for G and M are equal to the physical parameter values of the transmission channel. Due to the absence of local minima any adaptive method summarized in section 2.1.4 can be used for parameter estimation. However, using Newton's (and Newton's-like) or Levenberg-Marquardt methods the small gradients combined with high values (at the region $G > 5$ and $M < 1.5$) can cause convergence problems. This problem can be avoided by using simplex method.

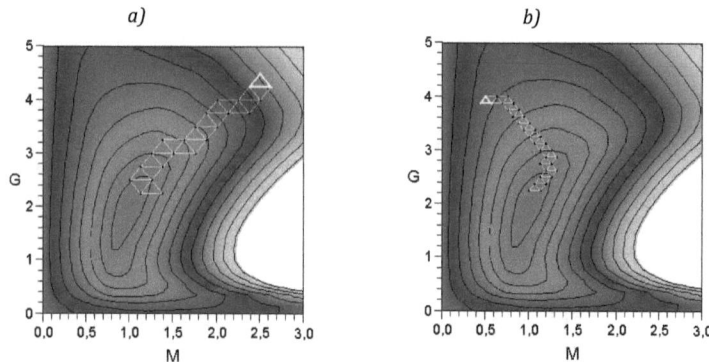

Figure 5.21: 3D learning function of the adaptive algorithm using simplex method with variations of the position of the first simplex and size of the simplex
a) first simplex with a side length of 0.50 and the center at $G = 4.50$, $M = 2.50$
b) first simplex with a side length of 0.25 and the center at $G = 4.00$, $M = 0.55$

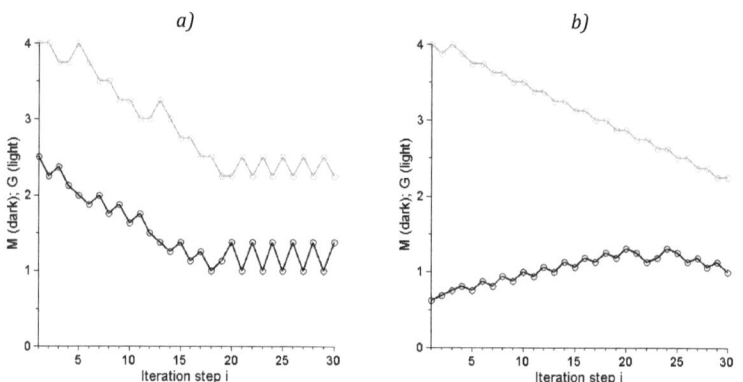

Figure 5.22: 2D learning function of the adaptive algorithm using simplex method with variations of the position of the first simplex and size of the simplex
a) first simplex with a side length of 0.50 and the center at $G = 4.50$, $M = 2.50$
b) first simplex with a side length of 0.25 and the center at $G = 4.00$, $M = 0.55$
G = air gap, M = magnetic dipole moment

The learning functions using simplex method are shown in Figure 5.21 and Figure 5.22 in three- and two-dimensional space, respectively. Steady state accuracy can be increased by using a smaller simplex or a variable simplex size (i.e., Nelder and Mead).

5.3.5 Complexity Estimation

For the complexity estimations of the adaptive algorithm, the summation of the squared difference (5.7) defined as error signal. The bit widths of the multi-bit signals are defined as 12.

One 12-bit subtractor and one 12-bit multiplier are required to calculate the squared difference and one 20-bit adder is required for summation. The maximal calculation frequency of the required arithmetic logic unit is equal to 900 kHz.

The simple physical model of the magnetic sensing arrangement is used to calculate the pulse shaped magnetic field, i.e., the 400 filter coefficients. With the simple physical model one addition, two subtractions and eight multiplications are required to calculate each one of the 400 coefficients. With one iteration step per revolution and a rotational speed of 9,000 rpm, the 400 coefficients must be calculated in 6.6 ms. Calculation can be carried out in time multiplex resulting in a calculation frequency of 60,606 Hz.

To implement the adaptive algorithm, four 12-bit adders and nine 12-bit multipliers are required. The maximal sampling rate of the arithmetic logic unit is equal to 900 kHz.

5.4 Clock Signal Generation

5.4.1 Constant Number of Samples per Symbol

So far a constant rotational speed resulting in a constant sampling frequency for the pDFE was assumed. In the general case, the rotational speed of the automotive combustion engine varies in the range from 0 to 9,000 rpm resulting in frequency variations of the magnetization pattern on the wheel (called »pattern frequency«) between 0 and 9,000 Hz on a wheel with 120 magnetic regions.

CHAPTER 5 – SOLUTION OF THE PROBLEM

Figure 5.23.a and Figure 5.23.b show the magnetic field at the position of the sensor element at a rotational speed of 2,500 and 5,000 rpm, respectively. The shapes of both waveforms are similar; however, a change in frequency appears.

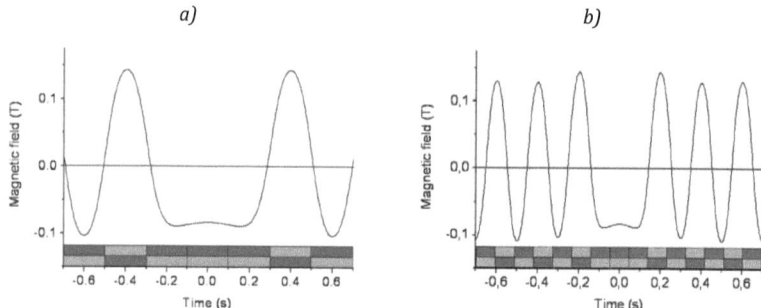

Figure 5.23: Magnetic field at different revolution speeds
Normal component on a gap of 1 mm and at a rotational speed of
a) 2,500 rpm
b) 5,000 rpm

As known from section 5.1, the magnetic field of a pole wheel can be calculated using linear superposition of the pulse shaped magnetic field of a single magnetic region. The width (expressed in time) of the pulse shaped magnetic field of a single magnetic region decreases with increasing rotational speed (Figure 5.24). Assuming no eddy currents, the pulse shaped magnetic fields shown in Figure 5.24 are equal but differently spread in time.

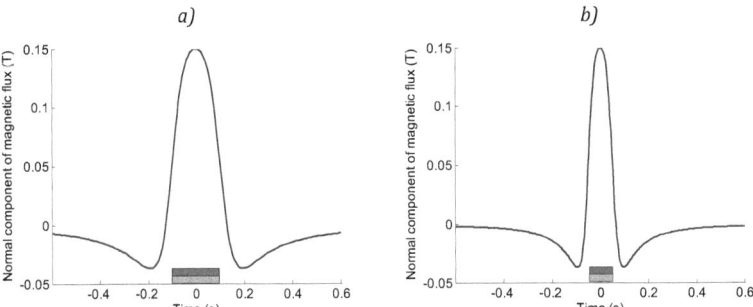

*Figure 5.24: Normal component of the pulse shaped magnetic field
generated by a single magnetic region
on a gap of 1 mm and at a rotational speed of
a) 2,500 rpm
b) 5,000 rpm*

Sampling the pulses with a constant frequency of 50 kHz yields the discrete samples shown in Figure 5.25. The number of samples per magnetic region—and so also the angle accuracy of the pDFE—varies with the revolution speed. With a constant sampling frequency for the pDFE, the filter coefficients have to change as a function of speed, which is difficult to implement.

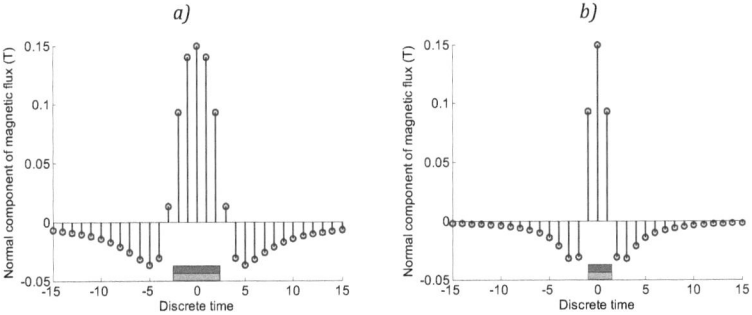

*Figure 5.25: Pulse shaped magnetic field of a single magnetic region
in the discrete time domain using a constant sampling frequency
at a rotational speed of
a) 2,500 rpm
b) 5,000 rpm*

Therefore a variable sampling frequency for the pDFE—having a constant number of samples per magnetic pole—is proposed. The sampling frequency of the pDFE has to

CHAPTER 5 – SOLUTION OF THE PROBLEM

follow the speed variations of the crankshaft. With this variable sampling frequency the filter coefficients do not need to be updated for speed variations (Figure 5.26).

In digital communication clock recovery techniques are commonly used to align the sampling frequency at the receiver with the sampling frequency at the sender [Bergmans'96]. Such techniques can only be used if the frequency varies within a narrow frequency band. Figure 5.27 shows two different data streams at different sampling frequencies and their corresponding analog waveforms. The analog waveforms of both digital signals are equal and therefore the sampling frequency cannot be extracted from the analog waveform. However, if the data sequence is known the sampling frequency can be extracted.

The pattern frequency in angle measurements has a wide frequency band (0 to 9,000 Hz) and clock recovery techniques from the area of digital communication cannot be used. The pattern frequency can only be determined from the analog waveform (Figure 5.23) if the pattern on the wheel (i.e., the data) is known. The modified PLL, presented in the following section, uses this approach and determines the rotational speed of the crankshaft.

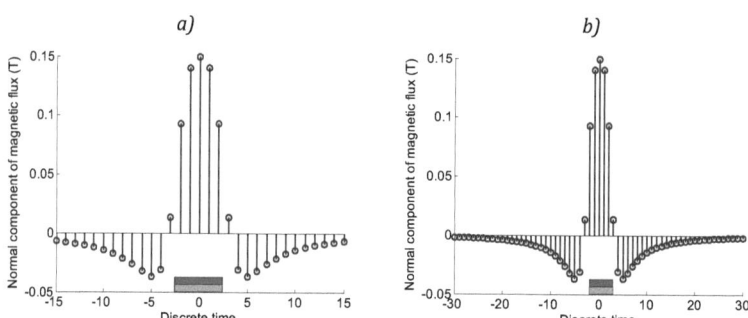

Figure 5.26: Pulse shaped magnetic field of a single magnetic region in the discrete time domain using a constant number of samples per magnetic region at a rotational speed of
a) 2,500 rpm
b) 5,000 rpm

Chapter 5 – Solution of the Problem

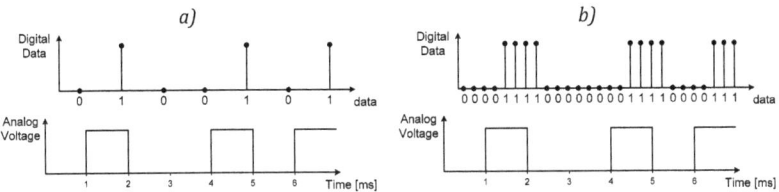

Figure 5.27: Digital data at different data rate and corresponding analog voltage
a) data rate of 1,000 Hz
b) data rate of 4,000 Hz

5.4.2 Crankshaft Speed Variations

A modified PLL structure can be used to align the sampling frequency $f_{\hat{k}}$ of the pDFE to the pattern frequency f_k of the rotating wheel (Figure 5.28). An analog and digital Schmitt trigger circuit are used to transform the analog signal $v(x)$ and the multi-bit signal $n[\hat{k}]$ into the binary signals $\theta_{v(x)}$ and $\theta_{n[\hat{k}]}$, respectively. The frequencies of the binary signals are proportional to the frequencies of their corresponding analog and multi-bit signal. The PLL structure compares the frequency (i.e., zero crossings) of the analog signal $v(x)$ with the frequency (i.e., zero crossings) of the calculated signal $n[\hat{k}]$. In the steady state the frequency of the calculated signal $n[\hat{k}]$ is equal to the frequency of the analog signal $v(x)$.

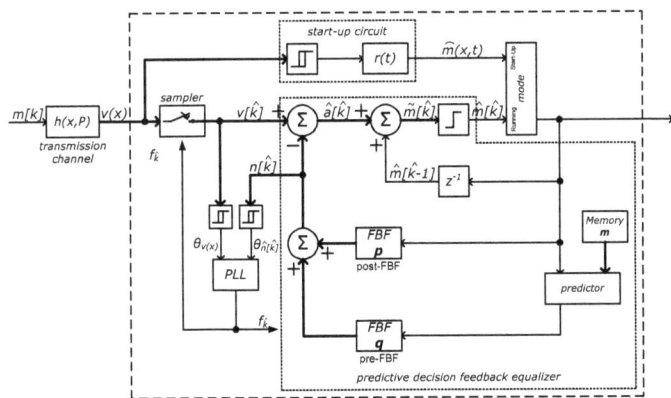

Figure 5.28: Predictive decision feedback equalizer with PLL

The closed loop of this modified PLL structure is shown in Figure 5.29. The structure is similar to a conventional PLL; however, the frequency divider is replaced by the pDFE.

In contrast to the ratio of a conventional clock divider, the ratio of the pDFE is not constant but depends on the pattern $\hat{m}_{\hat{k}}$.

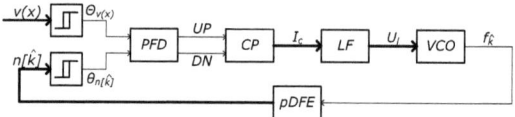

Figure 5.29: Digital phase locked loop used by the pDFE

To simplify calculations a regularly alternating magnetization pattern on the pole wheel was assumed. With a regular pattern the pDFE can be modeled by a clock divider with constant ratio (e.g., $\psi = 32$).

A digital PLL was designed as shown in [Best'04, Gupta'75] and a second order LF with the transfer function

$$F(s) = V_c \cdot \frac{s \cdot \tau_2 + 1}{s \cdot (s \cdot \tau_3 + 1)}, \qquad (5.8)$$

[Da Dalt'07] with $V_c = 1.82$ $1/\tau_2 = 200\,\text{Hz}$ and $1/\tau_3 = 2.2\,\text{kHz}$, was chosen. The frequency response of the loop filter was simulated using MATLAB and the result is shown in Figure 5.30.

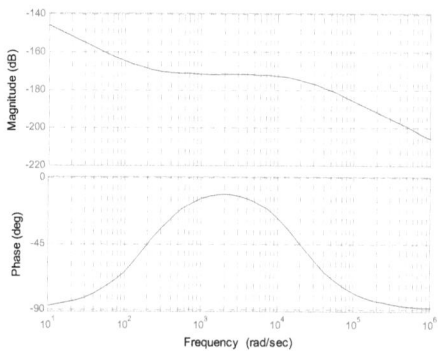

Figure 5.30: Bode diagram of the analog loop filter

Figure 5.31 shows the frequency and step response of the closed loop assuming that the PFD is working in the linear range between -2π and 2π. The output frequency of the PLL reaches its steady state value with an accuracy of 2% after 0.14 ms.

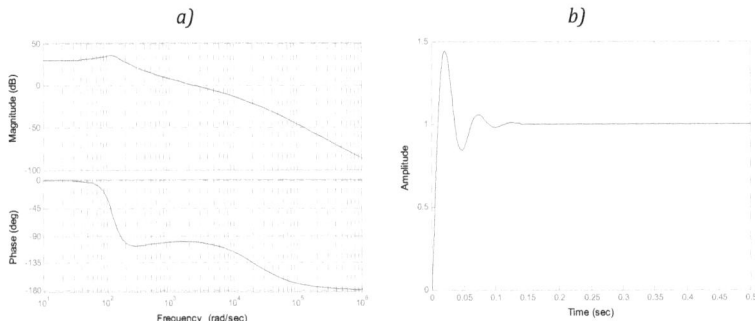

Figure 5.31: Characteristic of the closed phase locked loop
a) Bode diagram
b) step response

The performance of nowadays all-digital phase locked loops (ADPLLs) is similar or even better than the performance of analog PLLs [Kratyuk'07]. An additional benefit of the ADPLL is the higher degree of freedom in DLF and DCO design [Hsu'99, Kim'99, Xiu'04]. If a wide frequency range is required the effort of designing analog PLLs [Ahn'00] is higher than the effort of designing ADPLLs [Da Dalt'07]. Therefore the use of an ADPLL is proposed. The frequency range of an ADPLL is only limited by the bit-length of the counter and a frequency range of 450 Hz to 9,000 Hz and even more [Chung'03, Dunning'95] can easily be implemented.

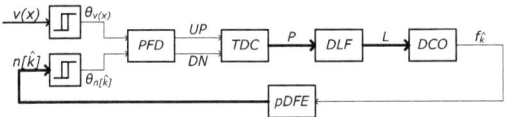

Figure 5.32: All-Digital PLL used by the pDFE

An all-digital PLL shows discretization and quantization errors. The discretization errors of the time to digital converter (TDC) were neglected because the length of the UP and DN pulses is much longer than the sampling time of the TDC [Kratyuk'07]. Also, the discretization effects of the digital loop filter (DLF, Figure 5.33) were neglected because a small sampling time of 1 µs is used for the TDC.

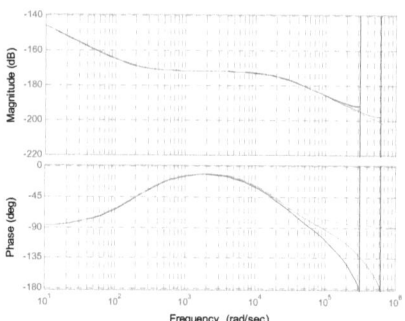

*Figure 5.33: Bode diagram of digital loop filter
with the sampling time as parameter
light dashed cyan: analog loop filter
dark blue: digital loop filter with $T_{lf} = 10$ μs
light green: digital loop filter with $T_{lf} = 5$ μs*

The output frequency f_k of the digitally controlled oscillator (DCO) is quantized to (2.16) and only integer scale down factors are possible. Pseudo decimal scale down factors can be realized by alternating the scale down factor between two integer values. For example, by alternating the scale down factor between 7.0 and 8.0, the pseudo decimal factor is equal to 7.5. The drawback of this solution is the increasing jitter at the PLL output.

5.4.3 Crankshaft Speed Fluctuations

5.4.3.1 Speed Fluctuations and Conventional Phase Locked Loops

The speed fluctuations of the crankshaft rotational speed result in frequency fluctuations of the pattern frequency f_k of the pole wheel (Figure 5.34). A conventional ADPLL cannot follow these frequency fluctuations and phase differences between the input (the pattern frequency) and the output frequency of the PLL appear [Hainz'08a].

CHAPTER 5 – SOLUTION OF THE PROBLEM

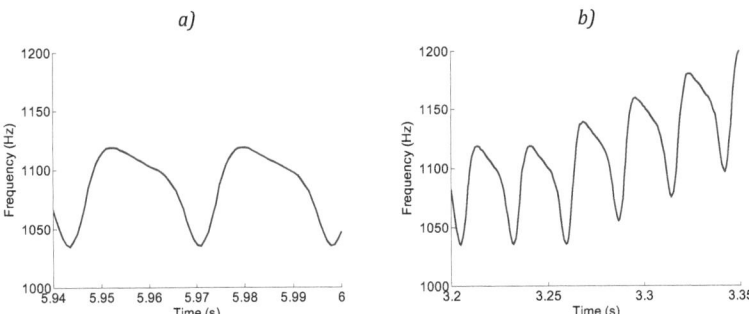

Figure 5.34: Frequency fluctuations of the pattern frequency
a) no speeding up the engine
b) speeding up the engine

Simulink simulations in the frequency domain were performed for conventional ADPLLs with different dynamical behavior. Figure 5.35 shows the simulation results using an ADPLL with fast (5 ms rise time) and slow (800 ms) step response of the closed loop. The slow ADPLL generates a constant output frequency and phase differences between $\theta_{v(x)}$ and $\theta_{n[k]}$ of up to 1.1 rad appear. Decreasing the rise time reduces the phase difference to below 0.7 rad. However, decreasing the rise time causes an increase in jitter at the output of the PLL [Da Dalt'07]. This makes clear that reducing the time constant is an insufficient strategy to minimize the phase difference between $\theta_{v(x)}$ and $\theta_{n[k]}$. A better solution for this problem is shown in the following section.

CHAPTER 5 – SOLUTION OF THE PROBLEM

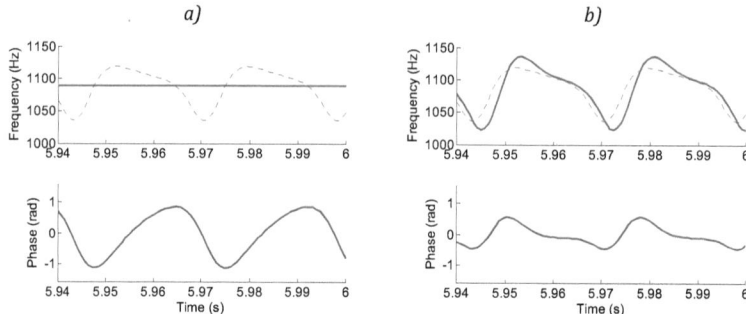

Figure 5.35 Simulation results of an ADPLL
dashed: input frequency fluctuations
solid: output frequency of the ADPLL (violet) and the corresponding phase error (red)
a) rise time of closed loop step response = 800 ms
b) rise time of closed loop step response = 5 ms

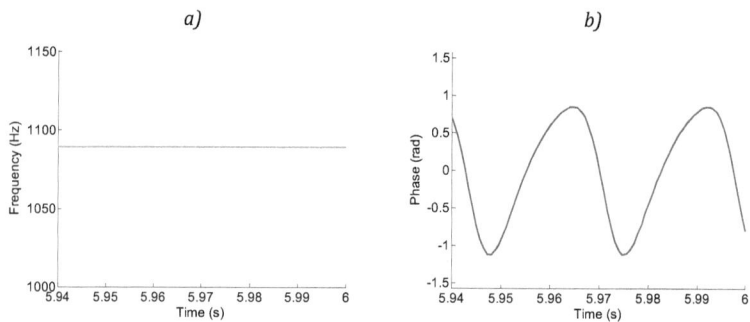

Figure 5.36: Frequency fluctuations expressed as constant frequency and phase shift
a) center frequency
b) phase shift

5.4.3.2 Speed Fluctuations and the Predictive All-Digital Phase Locked Loop

The frequency fluctuations at the input of the ADPLL can also be expressed by a constant center frequency and a varying phase shift (Figure 5.36). The proposed »predictive ADPLL« (pADPLL), shown in Figure 5.37, uses this approach [Hainz'08a]. A conventional ADPLL with a large time constant (compared to the frequency

-70-

fluctuations) generates the center frequency $f_{\hat{k}}$. The remaining phase difference between $\theta_{v(x)}$ and $\theta_{n[\hat{k}]}$ is measured and stored in a memory. The predictor estimates the next value for the phase difference by evaluating the last stored values. The variable delay line (VD) introduces this predicted phase difference into the clock signal $f_{\hat{k}}$. Assuming no prediction errors, the output signal $f_{\hat{k}'}$ is synchronous with the rotational speed of the engine crankshaft.

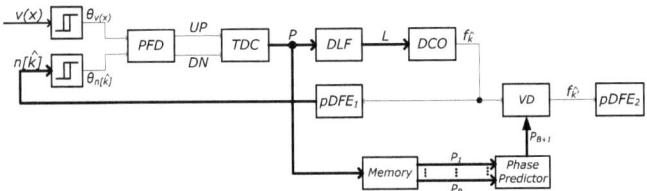

Figure 5.37: Predictive all-digital PLL

The innovative part of the pADPLL is the phase predictor. Different algorithms can be used to predict the next phase shift. One solution is to use polynomial interpolation of the past values and predict the next one using the analytic function of the interpolated values (extrapolation). Figure 5.38.a illustrates this prediction method interpolating the last two values with a first order polynomial.

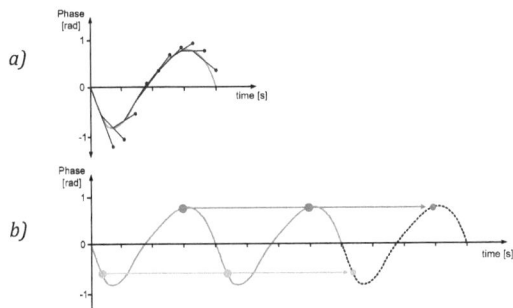

Figure 5.38: Phase prediction methods
a) polynomial prediction with linear interpolation of the last two values
b) periodic prediction with linear Interpolation of the last two cycles

Another predictive algorithm employs the periodicity of the engine speed fluctuations. At constant rotational speed the speed fluctuations—and consequently also the phase difference between $\theta_{v(x)}$ and $\theta_{n[\hat{k}]}$—are periodic. The algorithm stores the phase differences of the last engine cycles and reads them out for prediction. Polynomial

Chapter 5 – Solution of the Problem

interpolation or averaging at specific angle positions of the crankshaft as illustrated in Figure 5.38.b can also be used to improve the prediction accuracy. Typically the speed fluctuations differ slightly between the different cylinders of an engine. This can also be taken into account by storing and predicting the speed fluctuations of each cylinder separately.

Analyzing the crankshaft speed fluctuations is a commonly used technique to detect misfire events of a combustion engine (section 3.2.1). Using the pADPLL, misfire detection is possible on chip by evaluating the fluctuations of the phase difference.

Simulink simulations of a pADPLL in the frequency domain were carried out assuming a regularly alternating magnetization pattern along the circumference of the pole wheel. The fluctuating frequency, shown in Figure 5.34.a was used as input. An ADPLL with a large time constant of 300 ms generates the center frequency f_k and the phase difference between $\theta_{v(x)}$ and $\theta_{n[k]}$ remains (Figure 5.39.a). The predictor has a sampling time of 17 ms and uses second order polynomial interpolation of the last three stored values of the phase difference to estimate the next one. Small prediction errors appear if the slope of the phase difference changes (Figure 5.39.b). However, the main part of the remaining phase difference is caused by the finite sampling time of the predictor. This part can be decreased by using a higher sampling frequency or linear interpolation between the predicted values.

CHAPTER 5 – SOLUTION OF THE PROBLEM

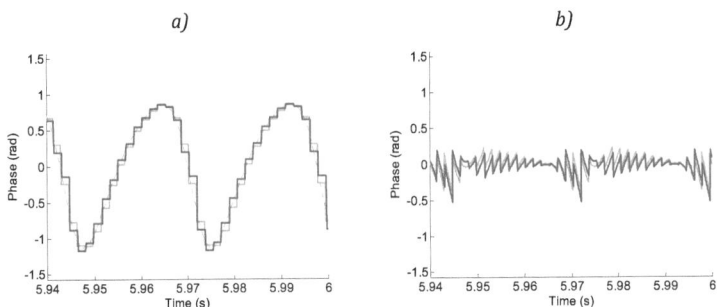

Figure 5.39: Simulation results of a pADPLL, no speeding up the engine
solid dark violet: predicted phase difference using polynomial prediction
solid light green: predicted phase difference using periodic prediction
dashed light red: simulated phase difference
a) phase difference between $\theta_{v(x)}$ and $\theta_{n[\hat{k}]}$
b) remaining phase difference after delay line

Because the frequency fluctuations at the input are periodic, no prediction errors appear if periodic prediction is used. The remaining phase difference (Figure 5.39.b) is only caused by the finite sampling time.

Compared to a conventional ADPLL with a small time constant (Figure 5.35.b), the phase difference between input frequency $\theta_{v(x)}$ and output clock $\theta_{n[\hat{k}]}$ was reduced from 0.75 rad to 0.45 rad using periodic prediction. Further improvement is possible by increasing the sampling frequency of the predictor or by using linear interpolation of the predicted values.

If the speed fluctuations are periodic, the results for phase prediction using periodic prediction are better than the results using polynomial prediction. On the other hand the results using polynomial prediction are better if non-periodic speed changes appear (e.g., speeding up the engine, Figure 5.40). Each algorithm is optimized for one mode of the engine (constant speed or speeding up). To enable highly accurate predictions at constant speed and during speeding up of the engine both prediction algorithms can be used in parallel. Then a decision unit is required to select the predicted value with the higher accuracy. Selection can be done by evaluating the accuracy of the last predicted value.

CHAPTER 5 – SOLUTION OF THE PROBLEM

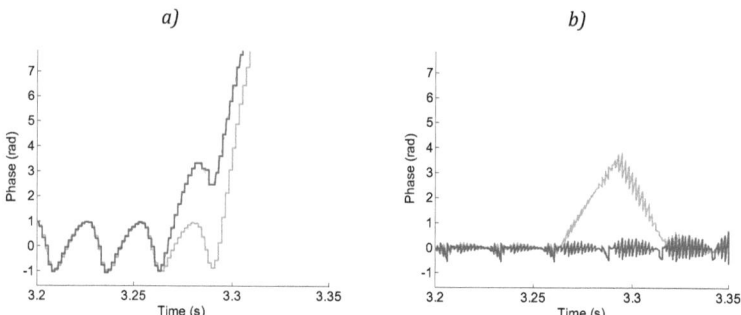

Figure 5.40: Simulation results of a pADPLL, speeding up of the engine
solid dark violet: predicted phase difference using polynomial prediction
solid light green: predicted phase difference using periodic prediction
dashed light red: simulated phase difference
a) phase difference between $\theta_{v(x)}$ and $\theta_{n[\hat{k}]}$
b) remaining phase difference after delay line

5.4.4 Implementation Hints

5.4.4.1 Aligning the Dynamical Behavior of the PLL to the Rev. Speed

A PLL with a dynamical behavior as shown in Figure 5.31 reaches its steady state value with an accuracy of 2% after 0.14 ms. During this 0.14 ms the crankshaft rotates 1 and 21 times at a rotational speed of 450 rpm and 9,000 rpm, respectively. The number of required revolutions to generate an accurate clock signal for the pDFE depends on the rotational speed of the crankshaft.

To make the number of required revolutions independent on the revolution speed, a speed dependent frequency characteristic of the DLF is proposed (Figure 5.41). A frequency meter (FM) measures the crankshaft rotational speed and the speed dependent filter coefficients of the DLF are read out from a LUT.

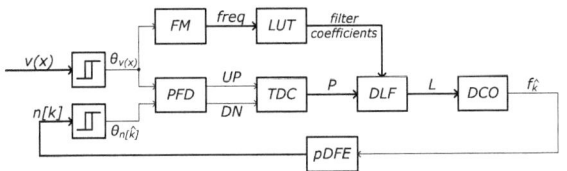

Figure 5.41: All-digital PLL with frequency dependent digital loop filter

5.4.4.2 Sampling frequency of the Variable Delay

Figure 5.42.a shows a typical implementation of a VD. The unit delays generate time-shifted versions of the input signal and the multiplexer selects one of these time-shifted signals (Figure 5.42.a).

In this application the delay line is used to introduce a phase shift between 0 and 2π to the output frequency $f_{\hat{k}}$ of the DCO and the frequency range of $f_{\hat{k}}$ is 450 Hz to 9,000 Hz. A phase shift of 1% (0.02π) at a frequency of 9,000 Hz is equal to a time shift of 1.1 µs. A phase shift of 99% (1.98π) at a frequency of 450 Hz is equal to a time shift of 2.2 ms. To realize a phase resolution of 1% over a frequency range of 450 Hz to 9,000 Hz, a delay line with a resolution of 1.1 µs and maximal delay of 2.2 ms is required. 2,000 unit delays and a 2,000×1 multiplexer are required to implement such a delay line.

Complexity can be reduced by clocking the VD with a variable sampling frequency (a multiple of the pattern frequency $f_{\hat{k}}$, Figure 5.42.b). With this variable sampling frequency the unit delays generate phase shifted versions of the input signal and the multiplexer selects one of these phase shifted signals. Using the structure with a variable clock frequency only 100 unit delays and a 100×1 multiplexer are required to realize a phase resolution of 1%. The variable clock with a multiple of the pattern frequency is generated by the PLL circuit.

CHAPTER 5 – SOLUTION OF THE PROBLEM

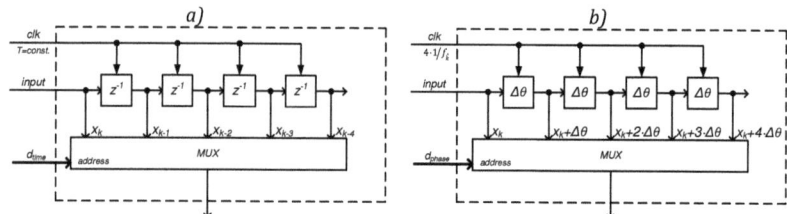

Figure 5.42: Implementation of a variable integer delay line
a) typical implementation using constant clock frequency (introduces delay)
b) implementation using a quadruple of the variable clock frequency (introduces phase shift)

5.4.4.3 Complexity reduction of the Variable Delay

The complexity of a 100×1 multiplexer is large and therefore an alternative implementation of a VD without of the need of a multiplexer is proposed (Figure 5.43).

The input of the VD is the frequency $f_{\tilde{k}}$ and the sampling frequency of the VD is $100 \times f_{\tilde{k}}$. Therefore the 100 unit delays of the VD store no random data but only the information of the transition point between two logical states of $f_{\tilde{k}}$. This time information can also be measured by a counter with high sampling frequency. If the counter starts counting with a rising edge of the clock signal, its counter value contains the time information of the rising edge.

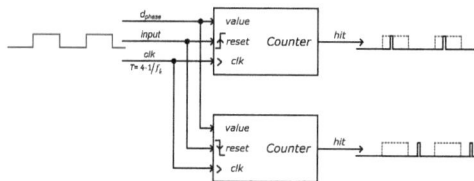

Figure 5.43: Implementation of a variable integer delay line using counters

Two counters are used to measure the time information of the falling and rising edge of the clock signal. If the counter value is equal to the value of the control signal d_{phase} (i.e., the wanted phase shift), its hit signal is logically high. The following circuit generates the delayed clock signal $f_{\tilde{k}'}$ using the hit signals of both counters.

5.4.5 Complexity Estimation

The phase resolution of the TDC was defined as $2\pi/100$ and therefore its maximal counter value is equal to 1,000. At a rotational speed of 9,000 rpm the sampling frequency of the TDC is equal to 90 MHz.

The rotational speed range was defined as 450 rpm to 9,000 rpm resulting in a frequency range of the DCO of 45 kHz to 900 kHz. Assuming the use of a high frequency clock signal of 90 MHz for the divide-by-N counter, its scale down factor varies between 2,000 and 100. High quantization error appears on a low revolution speed. At a rotational speed of 450 rpm the quantization error (i.e., Jitter) is equal to 1%.

The presented »predictive ADPLL« structure uses two digital filters: One of them (pDFE$_1$) tracks the center frequency $f_{\hat{k}}$ and the other one (pDFE$_2$) removes the ISI of the channel and generates the binary output signal. Therefore the hardware of one digital filter and a few computations are required additionally if the phase predictor is used. The pADPLL requires a VD. The phase resolution of the VD was defined as $2\pi/100$ and its maximal delay as 2π. The counter values of the VD counters vary between 0 and 100. At a rotational speed of 9,000 rpm the sampling frequency of the counters is equal to 900 kHz.

To implement the pADPLL one 10-bit, one 11-bit and one 7-bit counter are required. The highest frequency of the pADPLL is the reference frequency of the DCO which is defined as 90 MHz. The complexities of the PFD, DLF and phase predictor are negligible.

5.5 pDFE with Adaptive Algorithm and PLL

5.5.1 Simulation Results in the Time Domain

Simulations in the time domain can be carried out using state-driven (fixed simulation time steps) or event-driven simulators. To simulate the pDFE an event-driven simulator is recommended due to the varying clock frequency. Therefore, VHDL was preferred to Simulink for the simulations in the time domain.

CHAPTER 5 – SOLUTION OF THE PROBLEM

The modeled sensor arrangement consists of a rotating pole wheel with the pattern $m[c]$ (Figure 3.2.a) and GMR elements. The magnetic field at the position of the sensor element was obtained by FEM simulations presented in [Scherhäufl'06].

Four GMR elements are arranged in a Wheatstone bridge configuration (Figure 5.44.a). Two pairs of GMR elements are placed with a separation distance of 1 MR. The two elements of each pair are placed at 180° angle to each other. Therefore the elements R_{A1} and R_{A2} have a negative and positive sensitivity, respectively. Increasing the tangential component of the magnetic field at position x causes a decrease in the value of R_{A1} and an increase in the value of R_{A2} resulting in an increasing voltage $u(x)$.

GMR sensor elements with a resistance of $10,000\,\Omega$ and a sensitivity of $600\,\Omega/10\,\text{mT}$ were used for simulations (Figure 5.44.b).

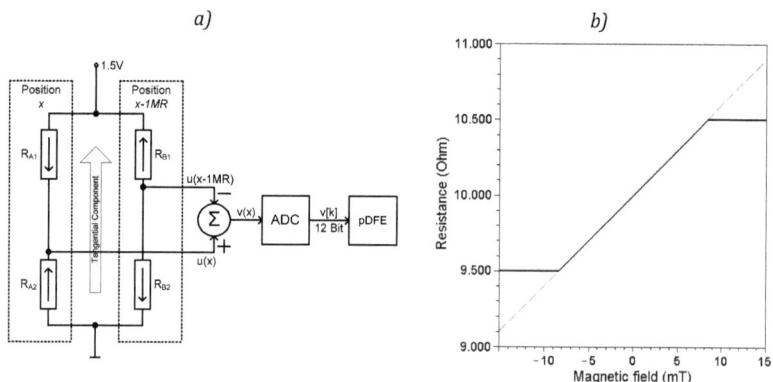

Figure 5.44: a) Four GMR elements in a Wheatstone bridge configuration
the arrows indicate the sensitive axis of the GMR elements
b) GMR resistance as a function of the applied magnetic field (positive sensitivity)
dashed light: no saturation; dark: with saturation

The pDFE structure, shown in Figure 5.45, was modeled as follows: the adaptive algorithm uses the simplex method to estimate the physical parameters P (air gap and magnetic dipole moment). The symbol density f_k was assumed to be by factor $\psi = 32$ higher than the symbol density f_c of the pattern $m[c]$ on the wheel. With this assumed higher symbol density the angle resolution is equal to $0.031\,\text{MR}$. The size of both FBF was defined as 144 filter coefficients which allows removing ISI over ±4.5 MR.

Chapter 5 – Solution of the Problem

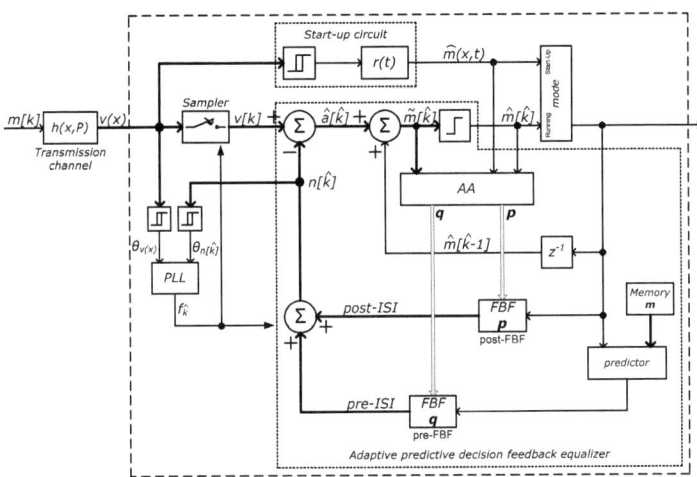

Figure 5.45: Predictive decision feedback equalizer with adaptive algorithm and phase locked loop

The start-up circuit measures the length between the zero crossings in signal $v(x)$ which allows detecting the synchronization point on the wheel. After the position is coarsely tracked a rough estimation $\bar{m}(x,t)$ of the pattern on the wheel can be generated. This rough estimation of the pattern is used to calculate the waveform $n[\hat{k}]$ using the simple model (Figure 5.46.a shows the calculated pre- and post-ISI). The parameter values are not yet estimated properly and therefore the calculated waveform using the simple model does not match the differential output voltage $v[k]$ of the Wheatstone bridge configuration. The discrepancies between these two waveforms identify remaining ISI and allow estimating the parameter values more accurately. The estimated physical parameter values converge to the physical parameter values of the channel.

A phase difference between the magnetic pattern $m[k]$ on the wheel and the signal $\hat{m}[\hat{k}]$ of the pDFE can be observed. The angle accuracy of the pDFE output signal is limited during start-up phase.

CHAPTER 5 – SOLUTION OF THE PROBLEM

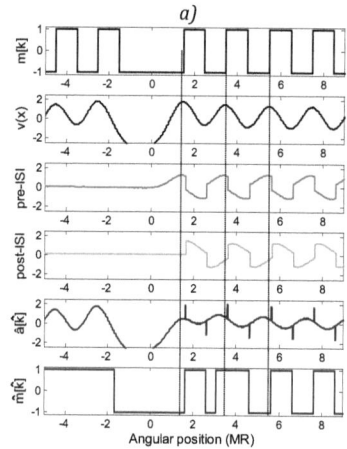
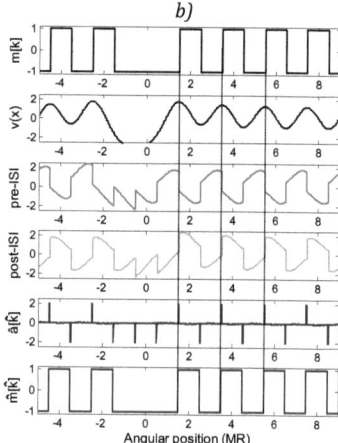

*Figure 5.46: Simulation results of the pDFE on a gap of 3.5 mm:
pattern on the pole wheel m[k], output voltage v(x),
pre- and post-ISI, internal signal â[k̂] digital output of pDFE m̂[k̂];
a) synchronization
b) adaption completed (after 20 iteration steps)*

After 10 iteration steps the pDFE switches into running phase where the signal $\hat{m}[\hat{k}]$ is fed into the feedback path instead of the signal $\hat{m}(x,t)$. The parameter values are estimated more accurately and the phase difference between $m[k]$ and $\hat{m}[\hat{k}]$ is reduced on each iteration step. With the well estimated parameter values also the calculated waveform $n[\hat{k}]$ matches the differential output voltage $v(x)$ of the Wheatstone bridge configuration (Figure 5.46.b shows the signals after 10 further iteration steps). The ISI of the channel is removed and the output $\hat{m}[\hat{k}]$ of the pDFE matches the pattern $m[k]$ on the wheel. A phase difference (i.e. angle error) of 0.07 MR between $\hat{m}[\hat{k}]$ and $m[k]$ remains. This value is higher than the resolution of 0.031 MR which is caused by the calculation errors of the pDFE (i.e., finite filter length of the pre- and post-FBF, finite adapting steps) and clock jitter.

The above simulations in the time domain have shown that ISI of a transmission channel can be removed by the pDFE. However, the main advantage of the pDFE is that the displacement of the zero crossing points can be compensated and the angle accuracy becomes independent of the gap between the sensor element and patterned wheel. Therefore simulations with different gaps were carried out as presented in [Hainz'08a]. ISI of the channels is removed also on larger gaps and also on larger gaps

CHAPTER 5 – SOLUTION OF THE PROBLEM

a phase error of 0.07 MR remains. The remaining phase error does not depend on the air gap.

5.5.2 Performance Limits

In the above section it was shown that the accuracy of the pDFE depends on design parameters such as the finite filter length of the FBF and clock jitter. As will be shown in this section the noise of the GMR elements and the ADC also affect the accuracy.

GMR elements have many well known noise sources such as Johnson-Nyquist[4] noise [Jury'02], $1/f$ noise, shot noise and Barkhausen noise [Sanders'96]. However, the complete understanding of GMR elements is still under development [Jury'02] and other noise sources such as magnetic fluctuation noise [Jury'02, Tsiantos'03, Zhang'02], resistance fluctuation noise [Stutzke'05] and noise caused by switching between magnetic stable states (eigenmodes) [Schrefl'07] are rarely known. Due to this lack of knowledge the amount of noise of the used GMR elements was not calculated but obtained by measurements [Infineon'06]. Measurements have shown that no significant $1/f$ noise appears and that the white noise is by a factor of 1.8 higher than the Johnson-Nyquist noise of a $10\,k\Omega$ resistor at room temperature. The measured white noise consists of Johnson-Nyquist noise, shot noise, magnetic fluctuation noise, resistance fluctuation noise and noise caused by switching between magnetic stable states.

The PLL uses zero crossing detection to convert the analog signal $v(x)$ into a binary sequence. Therefore noise at $v(x)$ causes jitter in the binary sequence. The amount of jitter depends on the noise amplitude and the gradient of the signal $v(x)$ at the position of the zero crossing. This gradient depends on the air gap resulting in a gap-dependent output jitter of the PLL. If an all-digital PLL is used, an additional jitter component appears, caused by the integer scale down factor of the DCO. An ADPLL with a high frequency clock of 25.6 MHz and an output frequency range between 25,600 Hz and 256,000 Hz was chosen which allows following the speed variations of the engine between 800 rpm and 8,000 rpm. To reduced the output jitter of the ADPLL a large time constant was used (such PLLs are called »clean-up PLLs«) [Da Dalt'07].

---

[4] *In literature the Johnson-Nyquist noise is also called thermal noise. However, the magnetic fluctuation noise is also another kind of thermal noise. To avoid mix up in this thesis the term Johnson-Nyquist noise is preferred to thermal noise.*

CHAPTER 5 – SOLUTION OF THE PROBLEM

The differential analog output voltage of the Wheatstone bridge configuration is converted into a digital signal (Figure 5.44.a). To calculate the quantization noise of the ADC, the use of a 12-bit successive approximation register ADC (SARADC) with a sampling frequency of 200 kHz was assumed.

GMR-noise, ADC-noise and PLL-jitter affect the adaptive algorithm error signal. If the summation of the squared difference (5.7) is used to calculate the error signal, the value of the error signal increases with noise and jitter. The statistical distribution of the different noise amplitudes and jitter values result in a statistical distribution of the error signal values. This statistical behavior affects the performance function and a 3σ band appears. Figure 5.47 shows a cross section of the three-dimensional performance function with its 3σ band. The effects of noise and jitter are very small which is caused by the high signal to noise ratio of GMR elements.

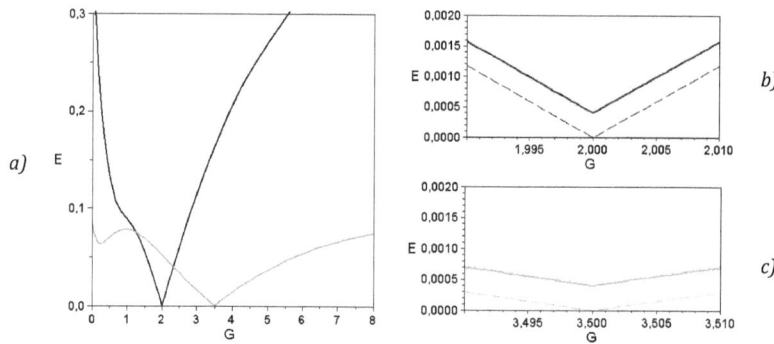

Figure 5.47: Cross section of the performance function
taking noise and jitter into account
solid: performance function with noise and its 3σ band
dashed: performance function with no noise
G = air gap, M = magnetic dipole moment, E = error signal

So far it was assumed that the pDFE is able to calculate the magnetic field without errors if the physical parameters are known. Due to finite filter order of the pre- and post-FBF, the length of the pulse shaped magnetic field is finite[5] (Figure 5.49.a). Calculating the magnetic field of the pole wheel by linear superposition of the finite

---

[5] Expressing the filter length as geometric length (MR) allows to make the calculations of the performance limits independent of the factor ψ. The required number of filter coefficients can be calculated by multiplying the filter length (expressed as geometric length) by the factor ψ.

pulses causes calculation errors as shown in Figure 5.48. Deviations between the measured and calculated waveforms appear even if the physical parameters are estimated correctly.

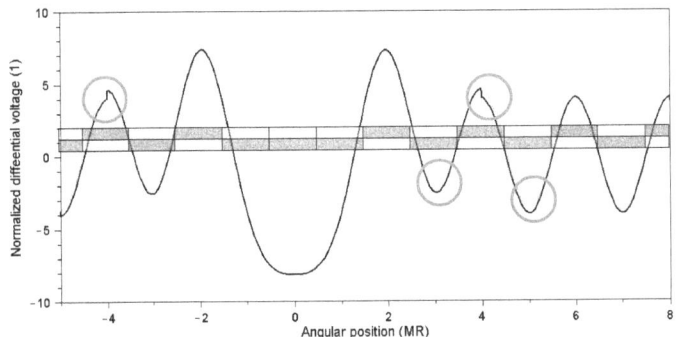

Figure 5.48: Calculated differential output voltage of a Wheatstone bridge on an air gap of 3.5 mm. The filter length of the pre- and post-FBF is infinite (light) and 4 MR (dark)

Figure 5.49 shows a cross section of the three-dimensional performance function where not only noise and jitter, but also the finite geometric filter length of the pre- and post-FBF was taken into account. Decreasing the geometric length of each filter from infinite to 4 MR and 2 MR causes an increase in the error signal. Discrepancies between the analog signal and the calculated waveform will remain even if the estimated parameter values match the parameter values of the channel. This affects the performance function and the local minimum is no longer at the exact position of the physical parameter values resulting in a wrong estimation result. The displacement of the global minimum and the increase in the error signal can also be seen in the three-dimensional performance function (Figure 5.50).

A filter length of the FBFs of 4 MR is required if the gap is below 3.5 mm. On gaps above 3.5 mm the use of larger filters is recommended.

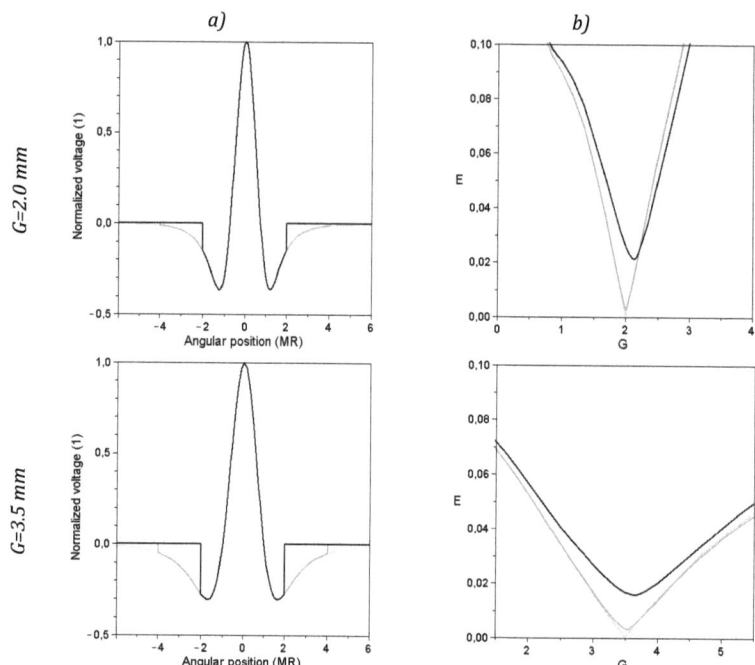

Figure 5.49: Finite filter length of the pre- and post-FBF
a) pulse shaped magnetic field with filter length of infinite (light), 4 MR and 2 MR (dark)
b) cross section of the performance function taking noise, jitter and filter length into account

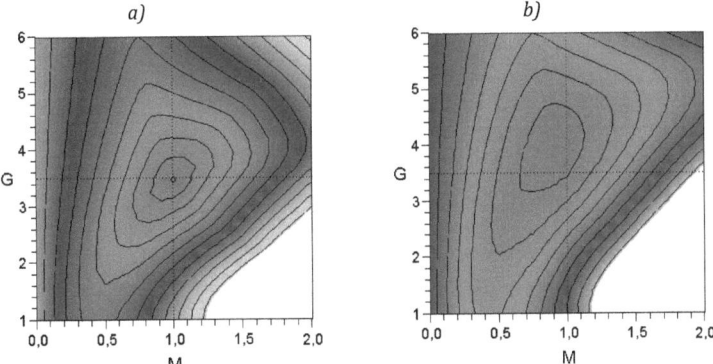

*Figure 5.50: Three-dimensional performance function
taking noise, jitter and the finite filter length into account;
The physical parameters are $M = 1.0$ and $G = 3.5$
a) infinite filter length
b) finite filter length of 2 MR*

CHAPTER 6

Conclusion

6.1 Summary

This thesis shows that high integration levels of today's semiconductor technologies allow improvements in the accuracy of measurements. The thesis presents a commonly used sensing arrangement for angle measurements based on magnetic technique. The angle accuracy of the presented arrangement is limited by a systematic error caused by variations of the air gap between the sensor element and pole wheel.

It is shown that the systematic angle error is caused by ISI. Therefore, the research analyzes known solutions to deal with ISI, for use in measurement applications. The presented solution uses a modified DFE to remove ISI and compensate the systematic angle error. The multi-bit forward filter of the conventional DFE is replaced by an additional single-bit filter in the feedback path. Compared to the conventional DFE, the modified DFE requires less hardware and does not introduce latency into the signal path.

The filter coefficients of the modified DFE are estimated adaptively. A physical model of the sensing arrangement is used and the physical values are estimated instead of estimating the filter coefficients directly. The presented structure estimates the air gap and the magnetic dipole moment of the pole wheel which allows measuring mechanical vibrations and aging of the wheel, respectively. Further benefits of using a physical model are the reduction in the complexity and the improvement in stability of the adaptive algorithm.

A simple model for the magnetic sensing arrangement using pole wheels is presented. The calculated magnetic field, using the simple model, is verified with FEM simulation results; it is shown that the deviation between them is below 2% for gaps between 1 mm and 6 mm.

Using a clock signal whose frequency is proportional to the angular speed of the crankshaft reduces the complexity of the DFE and adaptive algorithm. Because a conventional PLL cannot be used to follow the crankshaft speed fluctuations, a combination of a PLL and a variable delay line is proposed. The conventional PLL generates a clock signal with constant frequency, and the delay line introduces a variable phase shift which corresponds to the frequency fluctuations. Comparing the modified PLL to a conventional PLL in simulations, the phase error between the pattern frequency and the output frequency is reduced from 0.75 rad to 0.45 rad. As a further benefit, the crankshaft speed fluctuations are available and misfire events of the engine can be detected on sensor chip.

Compared to a conventional sensor without ISI cancellation, the systematic angle error is reduced in simulations from 1.8° to 0.45°. The angle accuracy is not limited by noise, but by the design parameters such as the filter length and clock frequency. Using today's semiconductor technologies, with the processing power and hardware available on a sensor chip, the angle accuracy can be reduced below the simulated value of 0.45°. The accuracy of the sensing arrangement can be defined by design and a tradeoff must be made between accuracy and complexity.

Complexity estimations for a filter structure which compensates the systematic angle error on air gaps between 1 mm and 4 mm are presented. The resolution of the structure is defined as 0.03° (i.e. 0.01 MR) and two single-bit FIR filters with 400 coefficients are used. To reduce complexity, the FIR filters are implemented as counters and LUTs.

As shown in the thesis, the start-up circuit and pADPLL are implemented by using six counters with bit-widths between 7 and 11. Additionally, a small arithmetic logic unit is required by the start-up circuit and the pADPLL to detect the synchronization point and to estimate the next value of the phase difference, respectively. Processing power is also required by the adaptive algorithm to calculate the 400 filter coefficients from the physical parameter values. For these calculations four adders and nine multipliers are required if multiplexing is used.

The complexity of a counter, adder or multiplier is small compared to the complexity of a 400×12 LUT. To give an example of the required chip size: To implement the

400×12 memory of the LUT in a 0.35 µm CMOS technology, an area of less than 0.35 mm² is required.

Besides complexity, also the maximal frequency of an electrical circuit plays a major role in the examination of its technical feasibility. At the presented implementation example, the signal with the highest frequency is the pADPLL reference clock of 90 MHz. This reference clock is scaled down by the DCO to a lower frequency between 45 kHz and 900 kHz which is the variable clock signal of the digital filter and adaptive algorithm.

Both, the area and frequency requirements of the presented filter structure are available using today's semiconductor technologies.

6.2 Outlook

Compared to sensor concepts employed today, the complexity of the presented structure is high. However, not all possibilities for complexity reduction are presented in this thesis and further reduction is possible. For example, the complexity of the memory and the calculation of the filter coefficients can be reduced by half if only every other coefficient is calculated and interpolation is used.

The presented phase predictor in the pADPLL uses extrapolation to predict the frequency fluctuations. A different approach is the use of a physical model of the engine and an adaptive algorithm. Using this approach, the physical parameter values of the engine such as torque and cylinder pressure are estimated by adaption. However, the engine speed fluctuations depend on many parameters and therefore a simple model might not be found. A second drawback of this approach is that with a physical model the predictor is restricted to a specific engine.

Due to the fast air gap variations caused by mechanical vibrations, the adaption is never completed and the estimated value of the air gap follows the physical value. Air gap variations caused by mechanical vibrations or a wobbling pole wheel are for the most part periodic. Storing the last values of the gap, the next value can be predicted by using extrapolation. The variances between the physical values and the adaptively estimated values of the parameters can be reduced using this approach.

Chapter 6 – Conclusion

In this thesis a simple model of the sensing arrangement using pole wheels was presented. To allow measurements also on toothed wheels, a simple model of the sensing arrangement using toothed wheels must be found.

References

[Ahn'00]   H. T. Ahn and D. J. Allstot, "A low-jitter 1.9-V CMOS PLL for UltraSPARC Microprocessor Applications," *IEEE Journal of Solid-State Circuits*, vol. 35, pp. 450-454, 2000.

[Alkhateeb'02] A. M. Alkhateeb and M. Das, "A robust algorithm for identifying different types of engine misfires," *The 2002 45th Midwest Symposium on Circuits and Systems*, 2002.

[Austin'67]   M. E. Austin, "Decision-Feedback Equalization for Digital Communication over Dispersive Channels," in *M.I.T.* Massachusetts: Cambridge University, 1967.

[AutoWeek'06]   AutoWeek, "Valeo signs up 'several global automakers' for camless engine," http://www.autoweek.com/apps/pbcs.dll/article?AID=/20061221/FREE/ /61218008/, 2006.

[Barry'04]   J. R. Barry, E. A. Lee, and D. G. Messerschmitt, *Digital Communication*, 3rd ed. New York, USA: Springer, 2004.

[Bergmans'96]   J. W. M. Bergmans, *Digital Baseband Transmission and Recording*. London, UK: Kluwer Academic Publisher, 1996.

[Best'04]   R. E. Best, *Phase-Locked Loops - Design, Simulation and Application*, 5th ed.: Mc Graw Hill, 2004.

[Bicking'94]   R. E. Bicking, "Magnetfeldsensoren," *Patent*, DE 44 35 678 A1, 1994.

[Blossfeld'05] L. Blossfeld, "Sensor mit Schwellenregeleinrichtung," *Patent*, DE 102 13 687 B4, 2005.

[Brown'96]   J. E. C. Brown, P. J. Hurst, and L. Der, "A 35 Mb/s mixed-signal decision-feedback equalizer for disk drives in 2-μm CMOS," *IEEE Journal of Solid-State Circuits*, vol. 31, pp. 1258-1266, 1996.

[Brown'99]   J. E. C. Brown, P. J. Hurst, B. C. Rothenberg, and S. H. Lewis, "A CMOS adaptive continuous-time forward equalizer, LPF, and RAM-DFEfor magnetic recording," *IEEE Journal of Solid-State Circuits*, vol. 34, pp. 162-169, 1999.

[Burden'00]   L. R. Burden and J. D. Faires, *Numerical Analysis*. n.s.: Brooks/Cole, 2000.

[Chung'03]   C.-C. Chung and C.-Y. Lee, "An all-digital phase-locked loop for high speed clock generation," *IEEE Journal of Solid-State Circuits*, vol. 38, pp. 347-351, 2003.

[Cioffi'90]   J. M. Cioffi, W. L. Abbott, H. K. Thapar, C. M. Melas, and K. D. Fisher, "Adaptive equalization in magnetic-disk storage channels," *IEEE Communications Magazine*, vol. 28, pp. 14-29, 1990.

[Da Dalt'07]   N. Da Dalt and C. Sandner, "PLL Design," *Course at Infineon Technologies Austria AG*, 2007.

[Deuflhard'04] P. Deuflhard, *Newton Methods for Nonlinear Problems. Affine Invariance and Adaptive Algorithms*. Berlin: Springer, 2004.

REFERENCES

[Draxelmayr'02] D. Draxelmayr and R. Borgschulze, "A mixed-signal Hall sensor IC with direction detection," *Proceedings of the European Solid-State Circuits Conference,* pp. 627-630, 2002.

[Dunning'95] J. Dunning, G. Garcia, J. Lundberg, and E. Nuckolls, "An All-Digital Phase-Locked with 50-cycle lock time suitable for high-performance microprocessors," *IEEE Journal of Solid-State Circuits,* vol. 30, pp. 412-422, 1995.

[Ellis'01] G. Ellis and J. O. Krah, "Observer-based resolver Conversion in industrial servo systems," *Power Conversion Intelligent Motion,* pp. 311-316, 2001.

[Fitzpatrick'07] R. Fitzpatrick, "Origin of Permanent Magnetism," http://farside.ph.utexas.edu/teaching/316/lectures/node77.html, 2007.

[Fleming'01] W. J. Fleming, "Overview of automotive sensors," *IEEE Sensors Journal,* vol. 1, pp. 296-308, 2001.

[Fricke'86] H. Fricke, H. Frohne, and P. Vaske, *Grundlagen der Elektrotechnik.* Stuttgart: B. G. Teubner, 1986.

[Götze'00] J. Götze, "Adaptive Filter: Theorie und Anwendung 1," *Lecture at Dortmund University of Technology,* 2000.

[Granig'07a] W. Granig, S. Hartmann, and B. Köppl, "Performance and technology comparison of GMR versus commonly used angle sensor principles for automotive applications, " *SAE Vehicle Sensors and Actuators,* pp. 1-14, 2007.

[Granig'07b] W. Granig, S. Hartmann, and B. Köppl, "Performance and technology comparison of GMR versus commonly used angle sensor principles for automotive applications," *Vehicle Sensors and Actuators,* pp. 1-14, 2007.

[Granig'07c] W. Granig, C. Kolle, D. Hammerschmidt, B. Schaffer, R. Borgschulze, C. Reidl, and J. Zimmer, "Integrated Gigant Magnetic Resistance based angle sensor," *IEEE Conference on Sensors 2006,* pp. 542-545, 2007.

[Gupta'75] S. C. Gupta, "Phase locked loops," *Proceedings of IEEE,* vol. 63, pp. 291-306, 1975.

[Hainz'08a] S. Hainz and D. Hammerschmidt, "Compensation of angular errors using Decision Feedback Equalizer approach," IEEE Sensors Journal, vol. 8, pp. 1548-1556, 2008.

[Hainz'06c] S. Hainz, D. Hammerschmidt, T. Werth, and H. Grünbacher, "Improving Phase Accuracy by removing systematic phase error introduced by Inter Symbol Interference," *Eurosensors,* vol. 1, pp. 1-4, 2006.

[Hainz'07a] S. Hainz and M. Jungwirth, "Magnetic Field Sensor Using a Physical Model to Pre-Calculate the Magnetic Field and to Remove Systematic Error due to Physical Parameters," *International Conference on Thermal, Mechanical and Multi-Physics Simulation Experiments in Microelectronics and Micro-Systems,* pp. 1-6, 2007.

[Hainz'04a] S. Hainz, E. Ofner, D. Hammerschmidt, D. Tatschl, and T. Werth, "Data predictive Decision Feedback Equalizer for position detection in automotive applications, " *IEEE International Conference on Industrial Technology,* vol. 1, pp. 1-4, 2004.

REFERENCES

[Hainz'04b]   S. Hainz, E. Ofner, D. Hammerschmidt, D. Tatschl, and T. Werth, "Position Detection by Inter Symbol Interference Removal for Engine Applications," *Austrochip*, pp. 137-140, 2004.

[Hainz'06d]   S. Hainz, E. Ofner, D. Hammerschmidt, T. Werth, and H. Grünbacher, "Position Detection in Automotive Application by Adaptive Inter Symbol Interference Removal," in *5th IEEE Conference on Sensors* South Korea, 2006.

[Halbo'80]   L. Halbo and J. Haraldsen, "The magnetic field sensitive transistor - a new sensor for crankshaft angle position," *SAE Congress and Exposition*, pp. 1-4, 1980.

[Hammerschmidt'07]   D. Hammerschmidt, "Introduction to signal processing," *Course at Infineon Technologies Austria AG*, 2007.

[Hammerschmidt'05]   D. Hammerschmidt, E. Katzmeier, D. Tatschl, W. Granig, J. Zimmer, B. Vogelgesang, and R. Rettig, "Giant magneto resistors - sensor technology & automotive application," *SAE World Congress*, pp. 1-16, 2005.

[Haykin'01]   S. Haykin, *Adaptive Filter Theory*. New Jersey: Prentice Hall, 2001.

[Hella'03]   K. G. Hella, "Technische Informationen - Elektronik - Kontaktlose Sensoren für X-By-Wire-Systeme," in *Hella - Inductive Sensors*, 2003, pp. 1-8.

[Hillier'04]   V. A. W. Hillier and P. Coombes, *Hillier's Fundamentals of Motor Vehicle Technology*. Cheltenham, UK: Nelson Thornes Ltd, 2004.

[Hobein'04]   D. Hobein, T. Dorißen, and K. Dürkopp, "Progress in automotive position sensors and introduction of the hella inductive position sensor," *SAE World Congress*, pp. 1-9, 2004.

[Hsu'99]   T. Y. Hsu, B. J. Shieh, and C. Y. Lee, "An All-Digital Phase-Locked (ADPLL)-based clock recovery circuit," *IEEE Journal of Solid-State Circuits*, vol. 34, pp. 1063-1073, 1999.

[Hu'03]   C. Hu, M. Meng, and P. X. Liu, "Microcomputer-based phase-discrimination capacitive angular sensor," *Proceedings 2003 IEEE International Symposium on Computational Intelligence in Robotics and Automation*, pp. 1439-1444, 2003.

[Hudzovic'01]   P. Hudzovic, *Optimalizácia*. Bratislave: Slovenská technická knižnica, 2001.

[Infineon'03]   Infineon, "TLE4925C," in *TLE4925C Datasheet*: Infineon Technologies Austria A. G., 2003.

[Infineon'05a]   Infineon, "Semiconductor Sensors," in *Semiconductor Sensors*: Infineon Technologies Austria A. G., 2005.

[Infineon'05b]   Infineon, "TLE4982C," in *TLE4982C Datasheet*: Infineon Technologies Austria A. G., 2005.

[Infineon'06]   Infineon, "Measurement results of a iGMR sensor element," Villach: Infineon Technologies Austria AG, 2006.

[Jurgen'99]   R. K. Jurgen, *Automotive Electronics Handbook*. New York: McGraw - Hill, 1999.

REFERENCES

[Jury'02] J. C. Jury, K. B. Klaassen, J. C. L. van Peppen, and S. X. Wang, "Measurement and analysis of noise sources in giant magnetoresistive sensors up to 6 GHz," *IEEE Transactions on Magnetics*, vol. 38, pp. 3545-3555, 2002.

[Kajley'97] R. S. Kajley, P. J. Hurst, and J. E. C. Brown, "A mixed-signal decision-feedback equalizer that uses a look-ahead architecture," *IEEE Journal of Solid-State Circuits*, vol. 32, pp. 450-459, 1997.

[Kim'99] N.-G. Kim and I.-J. Ha, "Design of ADPLL for both large lock-in range and good tracking performance," *IEEE Transaction On Circuit and Systems - II: Analog and Digital Signal Processing*, vol. 46, pp. 1192-1204, 1999.

[Kojima'04] T. Kojima, Y. Kikuchi, S. Seki, and H. Wakiwaka, "Study on high accuracy optical encoder with 30 bits," *The 8th IEEE International Workshop on Advanced Motion Control, 2004*, 2004.

[Kratyuk'07] V. Kratyuk, P. K. Hanumolu, M. Un-Ku, and K. Mayaram, "A design procedure for All-Digital Phase-Locked Loops based on a charge-pump Phase-Locked-Loop analogy," *IEEE Transactions on Circuits and Systems II: Express Briefs*, vol. 54, pp. 247-251, 2007.

[Le'02] M. Q. Le, P. J. Hurst, and J. P. Keane, "An adaptive analog noise-predictive decision-feedback equalizer," *IEEE Journal of Solid-State Circuits*, vol. 37, pp. 105-113, 2002.

[Lee'04] J. Lee, K. S. Kundert, and B. Razavi, "Analysis and modeling of bang-bang clock and data recovery circuits," *IEEE J. Solid-State Circuits*, vol. 39, pp. 1571-1580, 2004.

[Lequesne'98] B. Lequesne and T. Schröder, "Magnetic crankshaft and camshaft position sensor with a complementary geometry," *SAE International Congress and Exposition, Detroid*, pp. 1-9, 1998.

[Lequesne'99] B. Lequesne and T. Schroeder, "High-accuracy magnetic position encoder concept," *IEEE Transactions on Industry Applications*, vol. 35, pp. 568-576, 1999.

[Levenberg'44] K. Levenberg, "A Method for the Solution of Certain Non-Linear Problems in Least Squares," 1944.

[Madoglio'07] P. Madoglio, M. Zanuso, A. Levantino, C. Samori, and A. L. Lacaita, "Quantization effects in All-Digital Phase-Locked Loops," *IEEE Transactions on Circuits and Systems*, vol. 54, pp. 1120-1124, 2007.

[Marek'03] J. Marek, H. P. Trah, Y. Suzuki, and I. Yokomori, *Sensors for Automotive Technology*. Weinheim: Wiley-VCH Verlag GmbH, 2003.

[Marks'80] J. D. Marks and M. J. Sinko, "A Wiegand effect crankshaft position sensor," *SAE Congress and Exposition*, pp. 1-4, 1980.

[Mc Clellan'99] J. H. Mc Clellan, R. W. Schafer, and A. Yoder Mark, *DSP First a Multimedia Approach*. New Jersey, USA: Prentice Hall, 1999.

[Montani'08] M. Montani, N. Speciale, and N. Cavina, "Misfire Detection by a Wavelet based Analysis of Crankshaft Speed Fluctuation," *Circuits, Signals and Systems - CSS 2004*, 2008.

REFERENCES

[Moody'96]   K. L. Moody, R. Vig, K. P. Scheller, J. M. Towne, and T. L. Tu, "Detection of passing magnetic articles while periodically adapting detection thresholds to changing amplitudes of the magnetic field," *Patent*, US 5,650,719, 1996.

[Moody'97a] K. L. Moody, R. Vig, P. K. Scheller, J. M. Towne, and T. Tu, "Verfahren zum Erkennen von passierenden magnetischen Artikeln, bei dem die Erkennungsschwellenwerte periodisch an die sich verändernden Amplituden des Magnetfeldes angepaßt werden," *Patent*, DE 197 01 260 C2, 1997.

[Moody'97b] K. L. Moody, R. Vig, J. M. Towne, and T. Tu, "Verfahren zum Erkennen von passierenden magnetischen Artikeln mit einer Geschwindigkeit bis nahe Null," *Patent*, DE 197 01 261 A1, 1997.

[Motz'04]   M. Motz and T. Werth, "System zur Auswertung eines Sensorsignals,"

[Nelder'64] J. A. Nelder and R. Mead, "A simplex method for function minimization," *The Computer Journal*, 1964.

[Oberdorfer'61] G. Oberdorfer, *Die Wissenschaftlichen Grundlagen der Elektrotechnik*, 6th ed. München, D: R. Oldenbourg Verlag, 1961.

[Preis'04]   K. Preis, "Theorie der Elektrotechnik," *Lecture at Graz University of Technology*, 2004.

[Qureshi'82] S. Qureshi, "Adaptive equalization," *IEEE Communications Magazine*, vol. 20, pp. 9-16, 1982.

[Qureshi'85] S. U. H. Qureshi, "Adaptive equalization," *Proceedings of the IEEE*, vol. 73, pp. 1349-1387, 1985.

[Ramsden'06] E. Ramsden, *Theory and Application*, 2nd ed. Oxford, UK: Newnes, 2006.

[Rothenberg'97] B. C. Rothenberg, J. E. C. Brown, P. J. Hurst, and S. H. Lewis, "A mixed-signal RAM decision-feedback equalizer for disk drives," *IEEE Journal of Solid-State Circuits*, vol. 32, pp. 713-721, 1997.

[Sanders'96] S. C. Sanders, R. W. Cross, S. E. Russek, A. Roshko, and J. O. Oti, "Size effects and giant magnetoresistance in unannealed NiFe/Ag multilayers stripes," *Journal of Applied Physics*, vol. 79, pp. 6240-6242, 1996.

[Sauter'02] T. Sauter and N. Kerö, "System analysis of a fully-integrated capacitive angular sensor," *IEEE Transactions on Instrumentation and Measurement*, vol. 51, pp. 1328-1333, 2002.

[Sauter'05] T. Sauter, H. Nachtnebel, and N. Kerö, "A smart capacitive angle sensor," *IEEE Transactions on Industrial Informatics*, vol. 1, pp. 250-258, 2005.

[Scherhäufl'06] M. Scherhäufl, P. Fischer, and M. Jungwirth, "PDSSoC - Position Detection Sensor System on Chip," Upper Austria University of Applied Sciences, 2006.

[Schmidt'00] M. Schmidt, F. Kimmich, H. Straky, and R. Isermann, "Combustion supervision by evaluating the crankshaft speed and acceleration," *SAE World Congress*, pp. 1-8, 2000.

[Schrefl'07] T. Schrefl, "FEMME - Finite Element Micro MagnEtics," Vienna: SuessCo - IT consulting, 2007.

REFERENCES

[Spencer'99]   R. R. Spencer and P. J. Hurst, "Analog and mixed-signal circuits for digital communication," in *Final Report on 1999-00 MICRO*, 1999.

[Stutzke'05]   N. A. Stutzke, S. E. Russek, and D. P. Pappas, "Low-frequency noise measurements on commercial magnetoresistive magnetic field sensors," *Journal of Applied Physics*, 2005.

[Towne'01]   J. M. Towne, R. Vig, and K. P. Scheller, "Detection of passing magnetic articles with a peak-to-peak percentage threshold detector having a forcing circuit and automatic gain control " *Patent*, US 6,297,627, 2001.

[Treichler'01]   J. R. Treichler, C. R. Johnson Jr, and M. G. Larimore, *Theory and Design of Adaptive Filters*. New Jersey, USA: Prentice Hall, 2001.

[Tsiantos'03]   V. D. Tsiantos, T. Schrefl, W. Scholz, and J. Fidler, "Thermal magnetization noise in submicrometer spin valve sensors," *Journal of Applied Physics*, vol. 93, pp. 8576-8578, 2003.

[UoM'04]   UoM, "Speed fluctuations of a Diesel engine," www.eng.man.ac.uk/mech/merg/Research/React/4/4.html,2004, 2004.

[Valeo'08]   Valeo, "e-Valve System," www.valeo.com, 2008.

[Vig'04]   R. Vig, A. Biloti, and P. J. Bolsinger, "Magnetic gear tooth sensor with Hall cell detector," *Patent*, US 6,690,150 B2, 2004.

[Vig'94]   R. Vig, H. Yabusaki, and A. Bilotti, "Ein anstiegsaktivierter Hall-Spannungssensor," *Patent*, DE 694 25 465 T2, 1994.

[Walter'98]   W. L. Walter and A. P. Brokaw, "Magnetic field change detection circuitry having threshold establishing circuitry," *Patent*, US 6,064,199, 1998.

[Wang'07]   J. Wang and J. V. Sarlashkar, "Engine crankshaft position tracking algorithms applicable for given arbitrary cam- and crank-shaft position signal patterns," *SAE World Congress*, pp. 1-9, 2007.

[Widrow'85]   B. Widrow and S. D. Stearns, *Adaptive Signal Processing*. New Jersey, USA: Prentice Hall, 1985.

[Wolber'77]   W. G. Wolber, "Prime sensors for electronic automotive-engine control," *IEEE Transactions on Vehicular Technology*, vol. VT-26, pp. 144-150, 1977.

[Xiu'04]   L. Xiu, W. Li, J. Meiners, and R. Padakanti, "A novel all-digital PLL with software adaptive filter," *IEEE Journal of Solid State Circuits*, vol. 39, pp. 476-483, 2004.

[Zangl'02]   H. Zangl, S. P. Cermak, B. Brandstätter, G. Brasseur, and P. L. Fulmek, "Enhancing robustness of a new capacitive/magnetic full turn absolute angular sensor," *IEEE Instrumentation and Measurement*, pp. 129-133, 2002.

[Zhang'02]   J. Zhang, Y. Huai, and M. Lederman, "Analysis of magnetic noise in lead overlaid giant magnetoresistive read heads," *Journal of Applied Physics*, vol. 91, pp. 7285-7287, 2002.

Die VDM Verlagsservicegesellschaft sucht für wissenschaftliche Verlage abgeschlossene und herausragende

Dissertationen, Habilitationen, Diplomarbeiten, Master Theses, Magisterarbeiten usw.

für die kostenlose Publikation als Fachbuch.

Sie verfügen über eine Arbeit, die hohen inhaltlichen und formalen Ansprüchen genügt, und haben Interesse an einer honorarvergüteten Publikation?

Dann senden Sie bitte erste Informationen über sich und Ihre Arbeit per Email an *info@vdm-vsg.de*.

Sie erhalten kurzfristig unser Feedback!

VDM Verlagsservicegesellschaft mbH
Dudweiler Landstr. 99
D - 66123 Saarbrücken
www.vdm-vsg.de

Telefon +49 681 3720 174
Fax   +49 681 3720 1749

Die VDM Verlagsservicegesellschaft mbH vertritt

Printed by Books on Demand GmbH, Norderstedt / Germany